THE PROCESS OF FINE GRINDING

DEVELOPMENTS IN MINERAL SCIENCE AND ENGINEERING

VOLUME 1

THE PROCESS OF FINE GRINDING

BÉLA BEKE, D.SC. (ENG.)

1981

MARTINUS NIJHOFF / DR W. JUNK PUBLISHERS
THE HAGUE / BOSTON / LONDON

Distributors:

for the United States and Canada

Kluwer Boston, Inc.
190 Old Derby Street
Hingham, Ma 02043
USA

for all other countries

Kluwer Academic Publishers Group
Distribution Center
P. O. Box 322
3300 AH Dordrecht
The Netherlands

for Hungary, Albania, Bulgaria, China, Cuba, Czechoslovakia, German Democratic Republic, Democratic People's Republic of Korea, People's Republic of Mongolia, Poland, Roumania, Soviet Union, Socialist Republic of Vietnam, and Yugoslavia

Akadémiai Kiadó
P. O. B. 24
H-1363
Budapest
Hungary

This volume is listed in the Library of Congress Cataloging in Publication Data

ISBN-13:978-94-009-8260-4   e-ISBN-13:978-94-009-8258-1
DOI: 10.1007/978-94-009-8258-1

Joint edition published by
MARTINUS NIJHOFF/DR W. JUNK PUBLISHERS
P. O. B. 566, 2501 CN The Hague, The Netherlands
and
AKADÉMIAI KIADÓ
P. O. B. 24, H-1363, Budapest, Hungary

CONTENTS

PREFACE

Manuals of mineral dressing or more precisely those of comminution-classification treat in particular the mechanics of the machines, and generally their handling. In this way the plant engineer becomes acquainted with the equipment but is given no help in learning something of the processes taking place in the material to be comminuted even though the purpose of the operation is to enhance wanted and to avoid unwanted physical or physico-chemical processes.

Neglecting the description or representation of generally used and well-known equipment the object of this monograph is to supply information on the processes taking place in grinding installations. It explains the sometimes complicated phemonena by applying quite simple means; it requires only an elementary knowledge of mathematics, mechanics and physical chemistry.

The ideas are applicable to the grinding of all brittle raw materials or semi-finished industrial products. The special problems of cement grinding and those of ball mills are explained in more detail; in cement grinding we have to meet special requirements with regard to ball mills — apart from other considerations —, since these now demand the greatest overall energy consumption.

Currently disputed problems are dealt with, and naturally the views of the author are given in detail, but contrary views are also mentioned and the ample list of references ensures that these opposing views can be considered.

With regard to the measurement units, the newly introduced SI system was mainly used regardless of the fact that in the literature, most data — including empirical data — are in the kgf or in the English system; furthermore, such units are generally used in plant practice and could thus not be ignored. We therefore write tons instead of Mg, and microns instead of μm. Even the speed of rotation is given mostly in rpm and not in Hz. Where appropriate, symbols are given at the end of the chapter.

Dedicated to and primarily aimed at the plant engineer struggling with the problems of his everyday activity, the research worker will also find suggestions in this book for investigating actual problems.

CHAPTER 1

INTRODUCTION

Grinding is the fine phase of comminution or size reduction. In the coarse stage, that is, crushing, particles of mm–cm size are produced, the operation embraces two orders of magnitude. Grinding, in terms of equipment or product size, cannot be sharply distinguished from crushing; grinding produces particles of the micron–mm size embracing four orders of magnitude. Obviously processes of grinding are more complicated.

Comminution, which is an operation involving the application of mechanical energy, can be given three different definitions:

(a) The reduction of large, irregularly shaped solid particles to smaller sizes,
(b) The creation of new free surfaces,
(c) The changing of the number and size of the particles and surface of the mass. (These changes are connected with changes in the bond forces in the crystal lattice.)

The first definition seems to be self evident and is related to the coarse stage, to crushing.

The second definition results from the more than a hundred years old, but even today remarkable theory of Rittinger, it refers to the new contact-surfaces of vital importance in chemical reactions kinetics, it is characteristic to grinding.

The third definition characterizes the very fine stage of grinding. The requirement here is not the increase in heap surface area or particle number, nor the reduction in size, but that there is a change in whatever direction.

In the following, most of our considerations will be confined to processes in brittle materials which are unable to undergo lasting deformations; if the elasticity limit is exceeded, fracture will occur.

Before going into detail we shall summarize the three so to say "classical" theories of comminution.

(a) *Rittinger's "surface" theory* (1867) deals with comminution by imaginary slicing. The material, assumed to be homogeneous, of cubical form and of

9

x_1 cm in size, is to be sliced in the three main directions by $x_2 = x_1/v$ distanced parallel planes producing v^3 smaller cubes (Fig. 1/1). The ratio $v = x_1/x_2$ is the so called reduction ratio. According to the principle of Rittinger every individual slicing requires the same energy, i.e. the energy demand is proportional to the newly developed surfaces.

Starting from an initial $6x_1^2$, we get $6v^3 x_2^3$ final surface, the surface increase being equal

$$\Delta S = 6v^3 x_2^2 - 6x_1^2.$$

Relating this to unit volume we get the specific energy demand

$$W_1 = c\left(\frac{6v^3 x_2^2}{v^3 x_2^3} - \frac{6x_1^2}{x_1^3}\right) = c_1\left(\frac{1}{x_2} - \frac{1}{x_1}\right) \quad \text{J/cm}^3 \tag{1.1}$$

as the customary formula for the Rittinger theory.

If we write

$$W = cx_1^2(v - 1) \quad \text{J} \tag{1.1a}$$

it is clear that the energy demand increases with the growth of the reduction ratio.

In the frequent case when $v \gg 1$ or $x_1 \gg x_2$ the increase is proportional to the reduction ratio or inversely proportional to the product size

$$W_1 x_2 = c_1. \tag{1.1b}$$

The Rittinger formula calculates only the decomposing of molecular bond forces and neglects the work of elastic deformation preceding the fracture

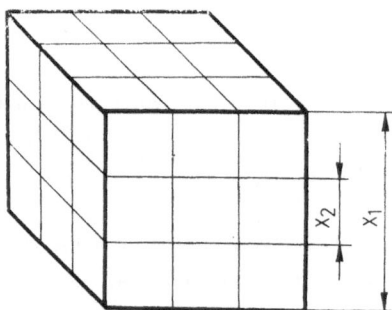

Fig. 1/1. Rittinger's principle

(b) *Kick's "volume" theory* (1885) takes even this deformation work into consideration. Accordingly, the energy demand for the fracture of a cube of size x will be

$$W = \int_0^\lambda x^2 \sigma \, d(x\lambda) = x^3 \int_0^\lambda \sigma \, d\lambda$$

where σ is the stress for fracture in kgf/cm² or Pa; $x^2\sigma$ the force for fracture; λ the specific deformation; $x\lambda$ the whole deformation. Accepting Hooke's proportionality law $\lambda = \sigma/E$ (E being Young's modulus, kgf/cm² or Pa), we get

$$W = \frac{x^3}{E} \int_0^\sigma \sigma d\sigma = x^3 \frac{\sigma^2}{2E} \ \text{J,} \tag{1.2}$$

that is, the energy demand is proportional to the volume of the body. Related to unit volume

$$W = \frac{\sigma^2}{2E} \ \text{J/cm}^3. \tag{1.2a}$$

The conspicuous defect of this formula is that it does not include the most important characteristic of the process, i.e. the reduction ratio. If the energy is supplied according to formula (1.2) fracture occurs and the body disintegrates to several particles of various sizes.

To eliminate this deficiency Bond and Wang (1950) suggested that consideration be given to the reduction of size x_1 to x_2 as the effect of multiple energy doses. If it is supposed that single energy dose results in an average reduction ratio ν_0 and the operation must be repeated z-times, then $\nu = \nu_0^z$ or $z = = \log \nu/\log \nu_0$ and the complete energy demand is

$$W_2 = zW = c_2 W \log \nu = c_2 \frac{\sigma^2}{2E} \log \frac{x_1}{x_2} \ \text{J/cm}^3. \tag{1.2b}$$

The most conspicuous difference of both theories manifests itself in the estimation of the progress of the comminuting process. If, for example, a mass of material is to be communited in succeeding steps from 1000 microns to 100, then from 100 to 10, thence from 10 to 1 micron, the energy-demand of the succeeding steps will be tenfold according to Rittinger, whereas according to Kick it remains unchanged; this latter is contrary to experience.

(c) The decades long debate was solved by *Bond* with his so called *third theory* (1952). This states that a particle of size x_1 has a total energy content W_{x1} necessary to reduce the size from being non-finite to x_1. This total energy can be calculated in two steps. The energy required for the first fracture is, according to Kick, proportional to x_1^3. Following the first crack the energy flow is directed to the surfaces, being proportional to x_1^2. So the energy-consumption must be between x_1^3 and x_1^2. Bond takes it to the arbitrary value of $x_1^{5/2}$. The total energy related to unit volume is $x_1^{5/2}/x_1^3 = 1/\sqrt{x_1}$. Size reduction from x_1 to x_2 will then require the difference of the total energies

$$W_3 = c_3 \left(\frac{1}{\sqrt{x_2}} - \frac{1}{\sqrt{x_1}} \right) \ \text{J/cm}^3. \tag{1.3}$$

If a geometrical interpretation is sought it can be seen that the energy demand is proportional to the square root of surface increment which has the dimension of length, which means a proportionality to new crack lengths.

Formula (1.3) is of course empirical too, but Bond proves its suitability for practical calculations by very many test results.

The total energy Wi of 1 sht (1 short ton = 907.2 kg) of material of 100 microns (called "work index" and regarded as an important material characteristic) can be calculated from the test results of size reduction from x_1 to x_2

$$\frac{Wi}{W} = \frac{\sqrt{\dfrac{1}{100}}}{\dfrac{1}{\sqrt{x_2}} - \dfrac{1}{\sqrt{x_1}}}$$

and

$$Wi = W \frac{\sqrt{x_1}}{\sqrt{x_1} - \sqrt{x_2}} \sqrt{\frac{x_2}{100}} \quad \text{kWh/sht} \tag{1.3a}$$

x substituted in microns.

The constancy of the work index for a given material demonstrates the applicability of the third theory.

In the knowledge of the work index the actual energy demand can be calculated as

$$W = 10 Wi \left(\frac{1}{\sqrt{x_2}} - \frac{1}{\sqrt{x_1}} \right) \quad \text{kWh/sht.} \tag{1.3b}$$

Of course neither the feed nor the product are composed of regular cubes or spheres so the meaning of size x must be clarified in all three theories. In industrial practice there are two arbitrary values in general use: x_{80} giving an undersize of 80% or $\bar{x}$ with the oversize of $100/e = 36.8\%$ (undersize 63.2%).

Symbols in Chapter 1
- x particle size, cm or μm
- v reduction ratio
- S surface, cm²
- W work or energy demand, J or mkgf, or J/cm³ or mkgf/cm³
- σ stress, kgf/cm² or Pa
- λ specific deformation
- E Young's modulus, kgf/cm² or Pa
- Wi work index, kWh/sht

CHAPTER 2

SINGLE PARTICLE BREAKAGE, ELEMENTS OF COMMINUTION PHYSICS

To acquaint ourselves with the complicated processes of comminution —which processes are influenced by many parameters — let us first investigate the effect of unit force on a single particle. This will be called single particle breakage in contrast to the collective processes in commercial operations when a multitude of forces affects a multitude of particles.

The half-a-century of research work into single particle breakage resulted in important findings.

To execute this research work highly complex laboratory equipment was utilized: the compressions of micron-sized particles were observed microscopically, high speed cinematographic techniques enabled the analysis of the fracture of plates by tension, bending or shock (with up to 500 000 exposures per s) and, of course, a working knowledge of mathematics and physics is indispensable.

Since many aspects of these research activities are not yet applicable to everyday plant operations, we shall confine to some fundamental principles.

The development of comminution physics can be attributed to several outstanding scientists but we should like to spotlight three important papers, viz. Griffith (1921), Smekal (1937), and Rumpf (1962).

Force and energy are the two physical conditions that promote fracture.

The force condition requires a local tension on the fracture front surpassing the molecular tension strength. (Brittle fracture occurs always in consequence of tension, even in the case of compressive load tensile stress takes place.) The order of magnitude of molecular tension strength is 10^4–10^5 kgf/cm^2 (10^3–10^4 MPa) surpassing by 2–3 orders of magnitude the practical strength of brittle materials.

The explanation is given by stress peaks brought about by microfissures always present. This is the so called defect location theory by Smekal (1937).

Figure 2/1 demonstrates a cut test body. In the angle point there takes place a stress increase according to the formula

$$\left(\frac{\sigma_{\text{mol}}}{\sigma} \right)^2 = \frac{l}{\varrho} \tag{2.1}$$

where ϱ can have the minimum value equal to the distance of neighbouring atoms, i.e. some 10^{-1} nm. The ratio $\sigma_{mol}/\sigma = 10^2$–$10^3$ refers to a minimal fissure length l of micron order so the force condition of fracture is the presence of fissure lengths of micron order (Rumpf, 1962).

The energy condition requires deformation work which is at least equal to the surface energy to ensure the propagation of the initial crack. Griffith

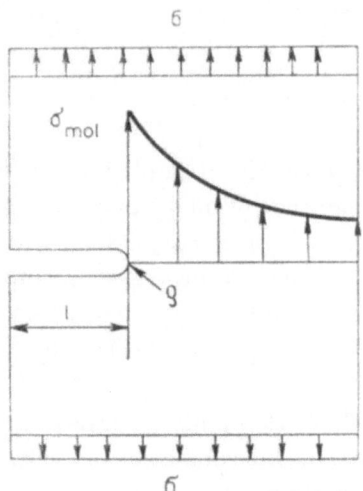

Fig. 2/1. Stress peak in a cut body (Rumpf 1962)

(1921) demonstrated that a minimum crack length, the so called Griffith length of micron order, is indispensable.

Griffith's fundamental equation is

$$-\frac{\partial U}{\partial l} \geq \frac{\partial S}{\partial l} \qquad (2.2)$$

where U is the work of elastic deformation and S the free surface energy. According to Griffith's theory, formula (2.2) holds only at a minimal initial crack length $l > l_{Griff.}$.

Rumpf (1962) proved that the circumstances are more complicated: propagation of the first fracture requires more energy or greater crack length than supposed by Griffith; other kinds of energy such as plastic deformations, structural and chemical changes (to be dealt with later), thermal effects, etc. also playing a role.

According to Rumpf and written in a simplified form

$$\frac{\partial S}{\partial l} = 2\gamma$$

and

$$G = -\frac{\partial U}{\partial l}$$

where γ is the free specific surface energy related to the $l \cdot dl$ differential surface (factor 2 for the two new surfaces), G the energy supplied to the elementary surface (known as "crack extension force"). Thus, the original Griffith equation gains the form

$$G \geq 2\gamma \qquad (2.2a)$$

and, marking the dynamic phase, the kinetic energy of the particle sprinkling as T, the energy condition becomes

$$\Sigma G = 2\gamma + \frac{\partial T}{\partial l} \qquad (2.3)$$

With the aforesaid ideas in mind, fracture of brittle materials occurs in succeeding phases:

(a) with an irregular shape, the corners crumble and the shape becomes more and more regular
(b) elastic deformation
(c) separation of particles
(d) cutting into pieces and dispersing of the particles.

The velocity of fracture propagation has been established as being about 1600–1700 m/s.

Schardin (1962), based on cinematographic analysis of the fracture phenomenon, established the succeeding phases of propagation to be

(a) radial cracks starting from compression waves
(b) radial cracks from other centres
(c) border cracks, surface waves
(d) refractions.

The fracture of crumbled, near regular-shaped bodies was investigated by Rumpf and collaborators (1967) by compressing quartz glass balls of the size 10–130 microns. Four types of fracture were established: annular, conical, radial, and internal cracks. A typical fracture development is shown in Fig. 2/2.

In this respect it is interesting to mention two new characteristics introduced by Matsui (Jimbo 1972): the crushing initiation index (CII) in $cm^3/kgfcm$ (or cm^3/J), and the crushing extension index (CEI) in $cm^2/kgfcm$ or cm^2/J.

The recent investigations of Gildemeister–Schönert (1975) demonstrated that elastic waves had no influence on the formation of the primary cracks. The dominant fracture phenomena are meridian cracks, the analysis gave no symptoms for any dynamical effects.

To conclude this chapter it cannot be ignored that below the micron order brittle materials do not crush but undergo permanent plastic deformation (Steier–Schönert 1971). This is the phenomenon of microplasticity to be treated later.

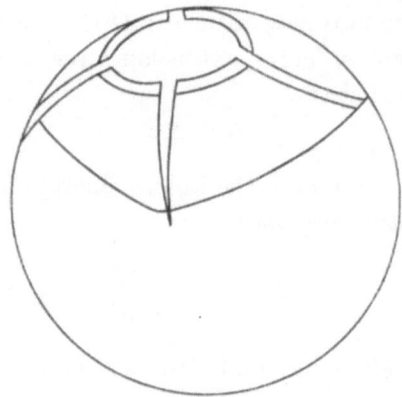

Fig. 2/2. Characteristic fracture of a sphere (Rumpf et al. 1967)

Symbols in Chapter 2
- σ stress, kgf/cm² or Pa
- l length, cm
- ϱ curvature radius, cm
- U elastic deformation work, J
- S surface energy, J
- γ free specific surface energy, J/cm
- G "crack extension force", J/cm
- T kinetic energy, J

CHAPTER 3

PARTICLE SIZE DISTRIBUTION
OF GROUND PRODUCTS

To enable us to get acquainted with industrial, bulk grinding operations let us begin with the description of the physical constitution of ground products characterized by their particle size distribution.

The mass of particles larger than about 40–60 microns can be determined by screening; below this range, determination is generally by sedimentation applying the Stokes formula.

In that the importance of particle size determination requires the elimination of manual work, various recording apparatuses were developed such as, for example, gravimetric registration of setting, signalling the change of transparency of suspensions (turbidimetry), microscopic particle counters with digital marking, electrical resistance change, light scattering in laser beams. To accelerate the speed of sedimentation in the size range below 10 microns, sedimentometers functioning in a centrifugal field were developed (e.g. as described in Németh–Horányi 1970).

Nowadays, because of automation, the on-line measurement of size distribution is gaining ground.

The whole problem of particle size analysis has been dealt with in detail in a set of papers by Leschonski et al. (1974–1975). Methods of on-line measurement are described by Davies (1973–1974).

The result is mostly a series of discrete values. Taking into account, however, the particle size x as a random variable we strive to describe the distribution in the form of a continuous function giving the ratios smaller than or greater than the size x. In this way we arrive at the $D(x)$ passing and $R(x)$ residue functions. The customary form of both functions is graphically presented in Fig. 3/1. A conspicuous factor is the dominating weight of the small particles characteristic for all ground products.

In grinding, in a seemingly simple operation, various forces act upon materials of the most diverse properties. Under these circumstances it will hardly be surprising that there should exist no simple formula of general validity for describing particle size distribution. Even so, many investigations

2

prove that the distribution can be described with an accuracy satisfying demands in practical use by functions containing two constant parameters.

It is, of course, well known that there are three functions which are applied in industrial practice.

In America the formula that is mostly applied is the one according to Gaudin and Schuhmann

$$D(x) = \left(\frac{x}{k}\right)^m. \tag{3.1}$$

In Europe the most often used function is that of Rosin and Rammler in its transcription by Bennett

$$1 - D(x) = R(x) = e^{-\left(\frac{x}{\bar{x}}\right)^n}. \tag{3.2}$$

or in its original form

$$R(x) = e^{-bx^n} \tag{3.2a}$$

Theoretical considerations based on suppositions not satisfying completely the real conditions led to the logarithmic normal function

$$D(x) = \Phi\left[\ln\left(\frac{x}{d}\right)^p\right] \tag{3.3}$$

where k, $\bar{x}$ and d are constant parameters of the dimension length representing a characteristic particle size, m, n and p are dimensionless constants related to the scatter of distribution, Φ the error integral (Gauss function).

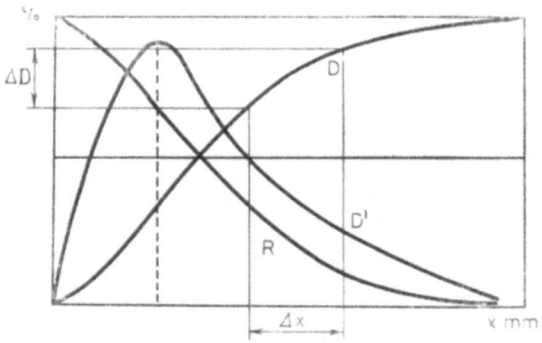

Fig. 3/1. Passing, residue and frequency curves

All three functions can be represented by straight lines applying a system of coordinates having a common abscissa ln x and the ordinates, respectively ln D, ln ln $1/(1 - D)$, and $\Phi(u)$, this last is called the logarithmic normal chart. The slope of the straight lines are m, n and p.

Our own experience and that of several research workers in the field prove that mostly the Rosin–Rammler formula can give the best approach to the particle size distribution by mass of ground industrial products, especially in cement manufacturing processes.

To compare the three kinds of graphical representation let us look at Fig. 3/2. Here the particle size distribution of a cement sample is presented with the abscissa ln x and ordinates as mentioned above. The best approach is given by the RR method. It is interesting to observe that the ordinate scale of the

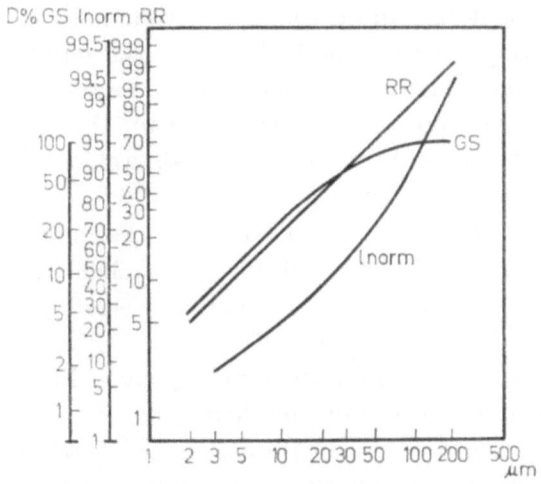

Fig. 3/2. Particle size distribution of a cement sample represented in GS, RR and lognormal chart

ognormal chart is most dense at and symmetrical to 50%; in the RR chart the scale is most dense at $100/e = 36.8\%$.

Before the mathematical analysis of the above distribution functions, it is necessary to introduce two concepts, viz. frequency function and specific surface.

The relative mass ΔD of the size fraction Δx can be read as the difference of ordinates (Fig. 3/1). Going to infinitesimally small Δx values we arrive at the $dD/dx = D'(x)$ frequency function giving the probability of the presence of particle size x (Fig. 3/1).

The surface of mass unity, known as specific surface, is frequently used to characterize the physical condition of ground products by one single value.

For simplicity, taking the particles as spheres or cubes, the specific surface can be calculated as

$$S = \frac{6}{\delta} \sum \frac{\Delta D}{x} \tag{3.4}$$

or in the knowledge of the distribution function

$$S = \frac{6}{\delta} \int_0^\infty \frac{D'(x)}{x}\, dx \tag{3.4a}$$

taking δ to be the density of the material in g/cm³, the specific surface is given in cm²/g.

For direct measurement of specific surface various methods are applied resulting in diverse values. The most widespread method is the measuring of air permeability (Blaine apparatus) having as its lower sensitivity limit 0.1–0.2 micron due to the thickness of the adsorbed air layer. The turbidimetry test (Wagner apparatus) has a sensitivity of micron order in accordance with the wavelength of light. The most accurate method is based on the adsorption of monomolecular layers (BET method) giving the real value excluding the surface of inner pores. This last method seems to be too complicated for plant laboratory practice. According to Anselm (1950), the surface values of two cement samples determined by various methods were

Wagner	Blaine	BET	Formula	(3.12)
1755	2655	6140	3590	cm²/g
1760	3185	7650	4250	cm²/g

Up to date investigations indicate much greater differences between the Blaine and the BET adsorption method (see, e.g. Opoczky and Mrakovics 1976). For scientific investigations, the BET adsorption method has been developed further—though with no consequences in plant practice. It has been stated that the result is dependent on the adsorbent medium, bigger molecules cannot penetrate into the pores. So, for example, nitrogen yields a smaller surface than water vapour.

If we return to the distribution functions, a mathematical comparison of the above three distribution formulae gives the following results.

The Gaudin–Schuhmann (GS) formula is an approximation of the one by Rosin and Rammler (3.2) expanded in series, and if only the first term is retained one arrives at (3.1). If the specific surface according to formula (3.4a) is calculated, the definite integral always has an infinite value in the case of GS, whereas in the case of RR: $n \leq 1$. The applicability of the GS formula is confined to the middle range of particle sizes, the straight line rises over 100% when $x > k$.

The theoretical deduction of formula (3.3) supposes—contrary to real circumstances—the homogeneity of the stochastic process. However, the probability of further size reduction decreases with smaller particle sizes; moreover, the quantity of bigger particles runs out because the supply from the above ceases. The consequence of this is the bending over of the curve given in Fig. 3/2. To eliminate the discrepancy from theory it was suggested by several scientists (e.g. Gebelein 1956, Fáy and Zselev 1963) that a fictitious simultaneous classification process be supposed at a transient, decreasing size limit. According to Gebelein a renormalized lognormal function plotted in the RR chart will be represented by a line of very small curvature having as an inflexion tangent the RR straight line.

If we combine formulae (3.3) and (3.4a) the specific surface can be calculated arriving, however, at a value far inferior to the real surface.

The above considerations result in the statement that all three distribution functions are mutual approximations of each other and the values m, n and p determining the slope of the straight lines can be accepted as nearly identical.

It thus seems to be justifiable to apply in our further considerations the RR formula. In view of this, it is obviously necessary to discuss it further.

At first sight its two constants are of no consequence, $\bar{x}$ being the size giving the sieve residue $100/e = 36.8\%$, n being the slope of the straight line in an arbitrary system of coordinates. (Both of these are quoted later as size module and uniformity coefficient, respectively.)

In terms of statistics, the distribution is characterized by a mean value M and standard deviation σ. Calculated for the RR function we get

$$M = \bar{x}\,\frac{1}{n}\,!\tag{3.5}$$

and

$$\sigma = \bar{x}\,\sqrt{\frac{2}{n}\,! - \left(\frac{1}{n}\,!\right)^2}\tag{3.6}$$

Another important statistical characteristic, the maximum place of frequency function or mode, is given by

$$x_m = \bar{x}\left(\frac{n-1}{n}\right)^{\frac{1}{n}}\tag{3.7}$$

having no real value if $n \leq 1$.

All these last three characteristics are so to say meaningless. But if we take as an argument the logarithm of particle size the situation becomes quite different. We can calculate the geometric mean of the distribution as (Beke 1964)

$$M_g = \bar{x}e^{-\frac{C}{n}}\tag{3.8}$$

where $C = 0.5772$, the Euler constant — , which is a more natural characteristic size of the distribution than the generally used $\bar{x}$ (size module) or x_{80}. More important, the standard deviation is equal

$$\sigma_l = \frac{\pi}{\sqrt{6}}\,\frac{1}{n} = \frac{1.282}{n}\tag{3.9}$$

index l denoting the logarithmic abscissa, i.e. the uniformity coefficient is inversely proportional to the standard deviation.

The mode is

$$x_{ml} = \bar{x}. \tag{3.10}$$

An important result is that if we use logarithmic abscissa, both constants of the RR distribution gain clear statistic significance. This result is hardly surprising in view of the fact that the use of logarithms provides a better simulation of the process bearing in mind the dividing and not substracting nature of comminution.

The above formula of standard deviation offers the possibility of determining the very important value of uniformity coefficient by calculation: the reading on the diagram is often very uncertain as the test points lie more or less scattered on both sides of the straight line. According to (3.9) and the concept of standard deviation (Beke 1970)

$$\frac{1}{n} = \frac{\sigma_l}{1.282} = 0.781 \sqrt{\Sigma \ln^2 x \Delta R - (\Sigma \ln x \Delta R)^2} \tag{3.9a}$$

where x and R are connected discrete values determined by laboratory methods.

In Table 1 the course of such a calculation is represented. The first row of x and R are the test results, the second row the fraction limit values, μ represents the logarithmic moment.

Table 1
Calculation of the uniformity coefficient

$x \ \mu m$	2	5	10	30	40	60	90
$x \ \mu m$	0.50	3.16	7.07	17.32	34.64	48.98	73.48	200
$\ln x$	−0.694	1.151	1.956	2.851	3.545	3.891	4.296	5.298
$\ln^2 x$	0.482	1.325	3.825	8.123	12.567	15.209	18.455	28
$R\%$	94	85	75	30	18	9	4	
$\Delta R\%$	6	9	10	45	12	9	5	4
$\ln x \Delta R$	−4.164	10.359	19.560	128.295	42.540	35.019	21.480	21.112
$\ln^2 x \Delta R$	2.892	11.925	38.250	365.535	150.804	136.881	92.275	112

$\Sigma \ln x \Delta R = \mu_1 = 2.742 \quad \mu_1^2 = 7.518$

$\Sigma \ln^2 x \Delta R = \mu_2 = 9.106$

$\sigma_l^2 = \mu_2 - \mu_1^2 = 1.588 \quad \sigma_l = 1.26$

$n = \dfrac{1.282}{1.260} = 1.02$

As mentioned above, the specific surface cannot be calculated starting from the RR distribution function. By deduction we arrive at the formula

$$S = \frac{6}{\delta} \frac{1}{\bar{x}} \Gamma \left(\frac{n-1}{n} \right) = \frac{6}{\delta \bar{x}} \frac{-1}{n}! \tag{3.11}$$

which is meaningless if $n \leq 1$, and which supplies no true values even if $n > 1$.

Anselm (1950) suggested for the range $n = 0.85$–1.2 the very simple empirical formula

$$S = \frac{36.8 \cdot 10^4}{\bar{x} n \delta} \quad \text{cm}^2/\text{g} \tag{3.12}$$

($\bar{x}$ to be substituted in micron, δ in g/cm^3).

It is interesting to note that the product $S\bar{x}$ (cm^2/cm^3) according to (3.11) is dependent solely on the uniformity coefficient n, but the arbitrary marginal scale of the diagram chart in the standard DIN 66145 sometimes furnishes values far from the true surface.

Concluding the problem of surface calculations, we cite here the formula by Bond (1961) based on the GS distribution formula (3.1)

$$S = \frac{60\,000m}{\delta k(1 - m)} \left[\left(\frac{k}{\text{Li}} \right)^{1-m} - 1 \right] \quad \text{cm}^2/\text{g} \tag{3.13}$$

The grind limit is Li with its value estimated to 0.1 micron. Formula (3.13) is to be employed for ground products of high scatter (wet grinding in ore dressing, m equal to about 0.5).

We give one more empirical formula (Kihlstedt 1962)

$$S = \frac{C}{\delta \sqrt{x_{80}}} \quad \text{cm}^2/\text{g} \tag{3.14}$$

C will vary from 500 to 1000 with the feed to be ground and the grinding method used. Formula (3.14) is applicable again for size distributions of high scatter.

Problems of classification and those of material mixtures will be treated later, but it is mentioned in advance that particle size distribution of classified or mixed ground products cannot be described by formulae dealt with in this chapter and presented as straight lines in the mentioned charts.

Symbols in Chapter 3

x	particle size, microns
$D(x)$	passing (undersize) function
$R(x)$	residue (oversize) function
k	particle size related to $D = 100\%$, microns
$\bar{x}$	particle size related to $R = 100/e = 36.8\%$, microns, quoted as "size module"
d	median particle size, microns
m, n, p	slope of distribution straight line
n	quoted as "uniformity coefficient"
Φ	error integral

$D'(x)$	frequency function
S	specific surface, cm^2/g
δ	density, g/cm^3
M	mean value of distribution, microns
σ	standard deviation, microns
x_m	mode, microns
μ	logarithmic moment
Li	grind limit, microns
Γ	gamma function

CHAPTER 4

KINETICS OF GRINDING

The progress of grinding as a function of time is described by the functions $D(x, t)$ or $R(x,t)$ with two variables. Many theoretical papers investigated this problem but, however, no practicable formulae for industrial use has resulted.

For illustration we confine ourselves to mentioning an equation of batch grinding by Austin and Klimpel (1964)

$$D(x,t) = D(x,0) + \int\limits_{0}^{t} \int\limits_{x}^{x_{max}} \frac{\partial D(y,t)}{\partial y} S(y) B(x,y) \, dy \, dt \qquad (4.1)$$

though of course this is of no practical use for plant operations [$D(x, 0)$ is the fraction by mass inferior to x; $S(y)$ the selection function for material of size y at time t; $B(x, y)$ the cumulative breakage distribution, i.e. the fraction finer than y in the breakage product of size x, and x maximum — the largest particle in the feed].

With some simplification of (4.1) Furuya et al. (1971) arrived at the formula

$$R(x, t) = R(x, 0) \, e^{-kx^n t} \qquad (4.2)$$

similar to the original RR function; formula (4.2) representing a first order of reaction for t, a frequent case of no general validity.

For practical calculations in plant operations we are compelled to be satisfied with the more simple $R(t)$, $x = $ const. function.

The semiempirical formula

$$R = R_0 e^{-ct^n} \qquad (4.3)$$

or

$$\frac{R}{R_0} = e^{-ct^n} \qquad (4.3a)$$

was proposed independently by Alyavdin (1938) (cited in Perov and Brand, 1954), and Chujo (1949) on the assumption that grinding velocity is the higher the more coarse fraction present in the mill, where R_0 is the sieve residue in the feed, t the grinding time (with ball mills approximately proportional to the distance from the feed end), c a constant connected mainly with the sieve mesh but with properties of the feed and that of grinding equipment too. Formula (4.3) is a mathematical consequence of the simultaneous validity of

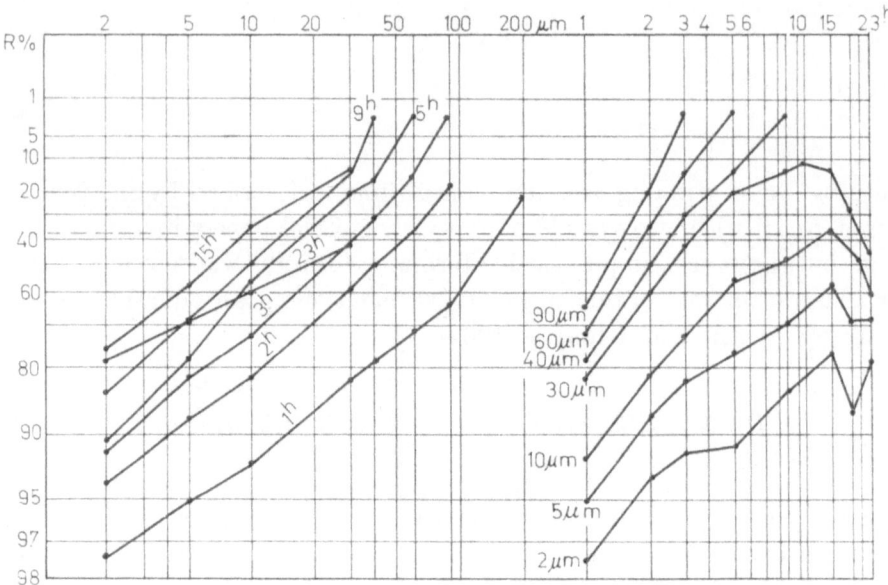

Fig. 4/1. Grinding of standard sand in laboratory ball mill. Sieve residue against sieve mesh (left) and grinding time (right). Representation in RR chart

the Rittinger and Rosin–Rammler laws. Therefore exponent n is equal to the uniformity coefficient in formula (3.2) or (3.2a).

Formulae (3.2a) and (4.3a) are formally identical: the change of sieve residue vs. grinding time can be represented in the RR chart—in the range of applicability of the Rittinger law—by straight lines too.

Figure 4/1 demonstrates the process of grinding standard sand in a laboratory ball mill, at the left side the sieve residue is plotted against sieve mesh, at the right side against grinding time. This figure enables us to make some important observations characterizing the grinding process.

After 5 hours grinding the lines $R(x)$ become more and more steep; after 15 hours the slope begins to flatten; after 23 hours simultaneously with the choking of the mill the line becomes abruptly flattened and simultaneously shifts to the right into the range of coarser particles: agglomeration of already ground particles appears.

26

The lines of $R(t)$ can be accepted as straight up to 15 hours in accordance with formula (4.3); later, the lines curve downwards signalling the gradual appearance of agglomeration which represents the limit of economic grinding.

By extensive investigations with diverse materials and grinding media these observations can be generalized. The uniformity coefficient, i.e. the slope of the lines, can be manifested as the function of a fraction, the numerator being the grinding time (t) or in the case of commercial ball mills with grinding media of different sizes the number of grinding impacts during the process, the denominator is connected with the energy of impacts (Beke 1964)

$$n = f\left(\frac{t}{Dd^p}\right) \tag{4.4}$$

or

$$n = f\left(\frac{N}{D^2G}\right) \tag{4.4a}$$

where D is the diameter of the ball mill shell, d the average diameter of grinding bodies, G their average weight, p having the value of about 6. Function (4.4a) for a clinker type is represented in Fig. 4/2; the downward curve (dashed) indicates agglomeration. Below certain particle sizes, the brittleness of materials comes to an end and we reach the state of microplasticity. As a consequence of further grinding impacts, a process analogous to hammer welding of metals takes place, bigger particles arise, the slope of the line of particle size distribution diminishes.

The results of investigations by Suzuki (1955) on a commercial multichamber 2.40×12 m cement grinding ball mill are in full agreement with those of the above laboratory scale experiments. At a nominal output of 16 t/h the uniformity coefficient attained the maximal value of about 1.08; at the lower output of 5 t/h (i.e. overgrinding) the value was 0.9 because of agglomeration; and at the highest output of 28 t/h, only 0.95 because of insufficient grinding impacts.

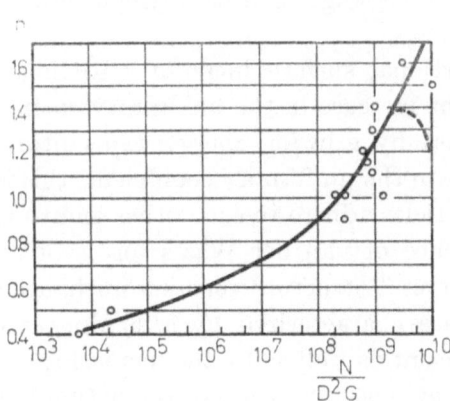

Fig. 4/2. Uniformity coefficient as function of number and individual energy of impacts

Formula (4.4) hints as to the possibility of influencing the value of the uniformity coefficient by changing the grinding body sizes. Agreement is demonstrated by the following experiment (see also Fig. 4/3). A commercial cement (d) was submitted for further grinding in a laboratory ball mill with identical energy supply but with different grinding bodies: 20 mm cylpebses (a), 8–12 mm steel balls (b), and 30–35 mm porcelain balls (c).

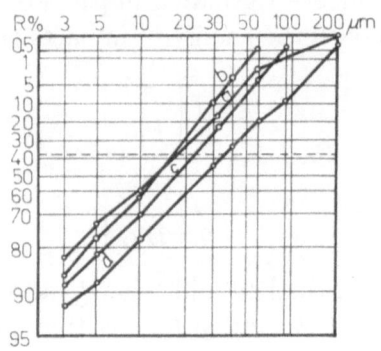

Fig. 4/3. Finish grinding of a cement sample (o)
(a) grinding media 20 mm cylpebs; (b) steel balls 8–12 mm; (c) porcelain balls 30–35 mm

The phenomena hindering fine grinding have three succeeding phases, viz. coating, aggregation, agglomeration.

In coating, because of unsaturated bond forces, particles adhere on the mill shell and grinding bodies. The free moving mass decreases, and cushioning impairs the efficiency of the grinding process. Coating can be eliminated by dosing with surface active agents.

As for aggregation and for agglomeration, a study of Fig. 4/1 is very instructive. Here again three stages can be distinguished (Opoczky 1977):

— energy proportional surface increase ("Rittinger phase") characterized by a slight increase in the uniformity coefficient,
— surface increase by growing energy expenditure characterized by a slight decrease in the uniformity coefficient; this is the stage of aggregation, and particles adhere to each other and to the grinding elements as a consequence of van der Waals forces of the magnitude 0.04–4 kJ/mol, the crystal structure remains unchanged,
— surface decrease characterized by a significant decrease in the uniformity coefficient as well as by the increasing size module; this is the stage of real agglomeration, where the acting bond forces are 40–400 kJ/mol; the crystal lattice suffers structural changes.

It should be noted that aggregation can take place following the grinding process in the conveying system or in storage bins. This detrimental phenomenon can be avoied by dosage of the grinding aids.

The agglomerated particles cannot be separated by screening, by sedimentation, or by the Blaine test. Aggregation and agglomeration have different behaviour in the course of the BET test. For agglomerated particles, surface active agents are ineffective.

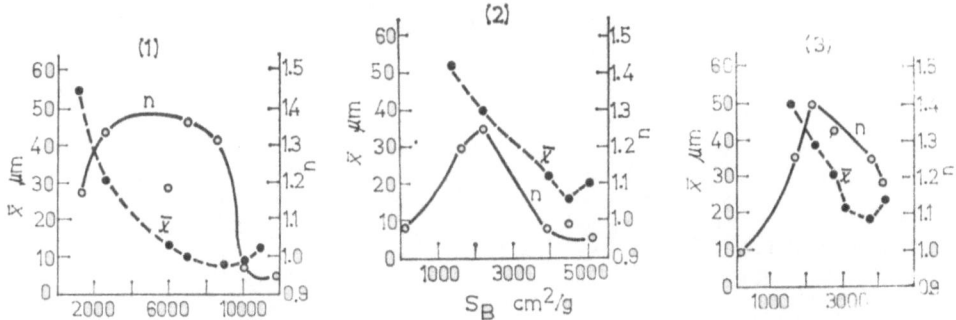

Fig. 4/4. Change in the characteristics n and $\bar{x}$ as a function of grinding fineness. (1) quartz; (2) C_2S; (3) cement clinker (Opoczky 1977)

To summarize, aggregation is signalized by decreasing uniformity coefficient n, agglomeration by increasing size module $\bar{x}$. The extreme value of n precedes that of $\bar{x}$. Some examples of these phenomena are presented in Fig. 4/4 (Opoczky 1977).

The mechanochemical relationships between these phenomena are discussed in Chapter 7.

Figure 4/5 presents these phenomena observed on clinker grinding in a laboratory ball mill. On the left side without, on the right side with dosing of surface active agents.

A formula characterizing the phenomenon of agglomeration was proposed by Papadakis (1960)

$$A = 1 - e^{-bDd^m} \tag{4.5}$$

where A is a factor indicating the rate lost by agglomeration from the energy-proportional surface increase according to Rittinger, b and m are constants.

Formulae (4.4) and (4.5) have the same character. In consequence:

— the limit of economical grinding is signalized by the flattening of the RR line,
— the greater shell diameter or greater grinding body size (i.e. bigger individual energy doses) promote agglomeration.

29

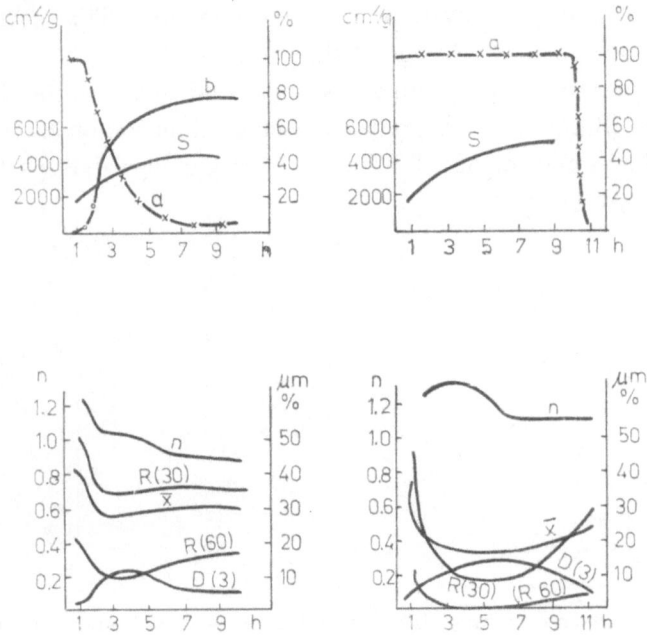

Fig. 4/5. Grinding process of Portland cement. Left: without, right with grinding aid
a free moving mass, %; *b* coated mass, %; *S* Blaine surface, cm²/g; $\bar{x}$ size module, μm; *n* uniformity coefficient *R*(60) and *R*(30) oversize to 60 and 30 microns, %; *D*(3) passing 3 microns, %

On differentiating formula (4.3) we get the function of grinding velocity, the amount of substance getting per time unit into the range smaller than the test sieve mesh:

$$\frac{dR}{dt} = - cnt^{n-1}R \tag{4.6}$$

having a constant value for $n = 1$, increasing with t grinding time when $n > 1$, decreasing when $n < 1$.

The uniformity coefficient can be regarded as a material characteristic (see Chapter 5), whereas c is dependent on the screen mesh and other conditions. According to experience, at constant sieve mesh, a larger n is connected to a smaller c.

Here, the concept of grinding equilibrium should be introduced; this phenomenon was first observed by Hüttig (1952). He ground electrolytic copper in a laboratory mill, the ground product was separated on a 60 micron sieve and both fractions were reground. The very interesting result was that owing to size reduction of the coarse, and the agglomeration of the fine fraction, the test resulted in both samples having an almost identical "equilibrium" particle size distribution. Hüttig repeated the test with brittle materials such as glass or marble with the same results; however, at finer granulometry. An important fact was established: brittleness is not an absolute, but a size-dependent material property.

To describe the kinetics of very fine grinding and grinding equilibrium starting from the differential equation

$$\frac{dS}{dW} = k(S_\infty - S)$$

Tanaka (1958) proposed the following formula

$$S = S_\infty(1 - e^{-kW}) \tag{4.7}$$

where W is the energy supply, S the specific surface, S_∞ that in the equilibrium state (Fig. 4/6). Because of surface decrease, in consequence of agglom-

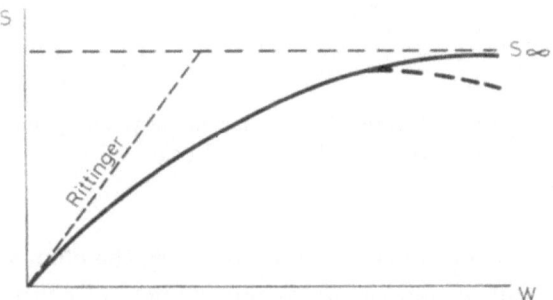

Fig. 4/6. Grinding limit according to Tanaka

eration, we have completed the figure by the dashed downward curving section.

Tanaka's formula was further developed by Harris (1967) in the form

$$S = S_\infty(1 - e^{-kW^m}) \tag{4.7a}$$

Formula (4.7a) also has no extreme value, the curve is inflexing and has a horizontal tangent at $W = 0$. It is interesting to note that formulae (4.7) and (4.7a) are both easily plotted in the RR chart as straight lines.

Jimbo and Suzuki (1973) proposed yet another variant in the differential equation

$$\frac{dS}{dW} = \left(\frac{dS}{dW}\right)_{t=0} - kS^m \tag{4.7b}$$

being valid past the Rittinger range as represented in Fig. 4/7.

Summarizing the above considerations, we can state: *agglomeration is brought about by energy oversupply* or high energy level. Consequently, high temperature promotes agglomeration, growing fineness requires often the cooling of the mill space.

The next task is to establish the connection between grinding fineness and mill output. In this calculation we suppose a constant energy supply which is self-evident with ball mills.

According to Rittinger's principle the surface increase is the countervalue of energy expenditure. In the case of bigger throughput, the same surface is dispersed on a bigger materials mass, the ground product will be coarser.

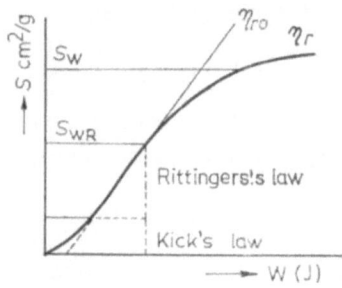

Fig. 4/7. Generalized representation of the relation between specific surface and work input, the effectivity cm²/J (Jimbo 1973)

With other grinding equipment, e.g. roller mills, the change in energy supply gives rise to a roughly proportional output change, of course with unchanged grinding fineness.

Starting from formula (4.3), according to Alyavdin (1938), it can be deduced, that if P_{10} (t/h) is the output with 10% residue on a random mesh sieve, then with R% residue the output will be $P = qP_{10}$ and

$$q = \frac{1}{\sqrt[n]{\log \dfrac{R_0}{R}}}. \tag{4.8}$$

where R_0 is the sieve residue of the feed, n the uniformity coefficient. In the frequent case of $R_0 = 100\%$ and $n = 1$ the formula gets the more simple form

$$q = \frac{1}{\log \dfrac{100}{R}}. \tag{4.8a}$$

Values of q according to (4.8) and $R_0 = 100\%$, are given in Table 2.

It is quite clear that in the case of smaller uniformity coefficient (or greater scatter), the output is more effectively influenced by changes in grinding fineness.

Supposing the simultaneous validity of the Rittinger and Rosin–Rammler formulae, the connection between P (t/h) output, sieve mesh h micron, and

Table 2

Values of output factor q

R%	1	2	3	4	5	6	8	10	12	15	20	50
$n = 0.7$	0.37	0.47	0.55	0.62	0.68	0.75	0.87	1	1.11	1.31	1.67	5.5
$n = 1$	0.50	0.59	0.66	0.72	0.77	0.82	0.91	1	1.08	1.21	1.43	3.3
$n = 1.2$	0.56	0.64	0.70	0.75	0.80	0.85	0.92	1	1.06	1.17	1.35	2.7

sieve residue R, can be deduced resulting in the formula

$$R = e^{-\left(\frac{ch}{P}\right)^n} \tag{4.9}$$

The units of c are t/h, µm, and can be calculated from a set of connected test values.

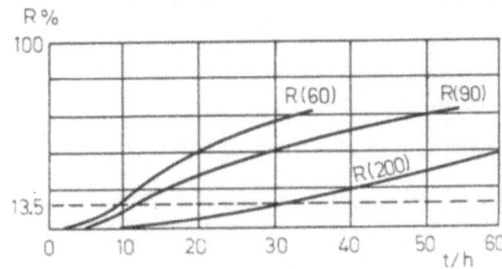

Fig. 4/8. Change of sieve residue against growing throughput

Formula (4.9) is equivalent with formula (4.8) and is represented in Fig. 4/8. To conclude this chapter, the unchanged uniformity coefficient presupposing the conversion formula from one sieve residue $R(h)$ to $R(h_1)$ is

$$R(h_1) = R(h)^{\left(\frac{h_1}{h}\right)^n} \tag{4.10}$$

or applied to the 90 micron sieve

$$R(h) = R(90)^{\left(\frac{h}{90}\right)^n} \tag{4.10a}$$

as easily follows from the original Rosin–Rammler formula (3.2).

Symbols in Chapter 4

t	grinding time, hours
y and x	particle size, microns
$R(t), R(x, t)$	residue (oversize) function
$D(x, t)$	passing (undersize) function
$B(x, y)$	breakage function

$S(y)$	selection function
R_0	sieve residue in the feed
n	uniformity coefficient
b, c, m, k, p	constants
D	mill shell inside diameter, m
d	grinding body size, m
N	number of grinding impacts
G	average mass or weight of grinding bodies, kg
A	coefficient of agglomeration
S	specific surface, cm²/g
W	energy input, J or kWh
P	output, t/h
h	sieve mesh, microns
q	output factor
c	constant, t/h, μm, in equation (4.9)

CHAPTER 5

GRINDABILITY OF MATERIALS

Grindability as a material characteristic must indicate the suitability for size reduction. Its value has to give the result of the grinding operation as a result of unit energy expenditure. And here we meet a difficulty: the result of the operation is a new particle size distribution which cannot be described by a single value. Therefore the grindability cannot be charaterized as a difference or quotient of two numbers.

In industrial practice, however, as a convention according to Lehmann and Haese (1955) grindability is defined as the relation of specific surface increase (cm²/g, determined using the Blaine test) to energy expense (J/g of kWh/g). The customary units of measure for grindability are cm^2/J or cm^2/kWh.

Grinding resistance is the reciprocal of grindability, the customary unit of measure being erg/cm^2.

As discussed in Chapter 3, the Blaine surface is sometimes far from the real surface. This circumstance gives, however, no rise to confusion in calculating grindability if the grinding fineness is characterized by Blaine surface too.

But in the very fine phase of grinding the above definition becomes meaningless; in the case of agglomeration grindability will have a negative value!

Generally speaking, grindability is not an absolute but a fineness dependent characteristic. With growing fineness, grindability decreases. Data in manuals are related to moderate grinding fineness, e.g. 10% residue on 0.09 mm sieve or 3000 cm²/g Blaine surface.

Industrial practice therefore simplifies further the problem. For a known grinding process (e.g. cement grinding in a ball mill to a grinding fineness of 3000 cm²/g Blaine), the grindability of a material is characterized in kWh/t, the grinding resistance in kg/kWh.

It is almost superfluous to mention that the grinding equipment influences these values too. So, for example, for cement raw meal grinding, different values will be obtained from grinding in a ball mill, in an autogeneous mill, or in a roller mill.

To determine the kWh/t material characteristics, taking into account the above limitations, there are several test methods applied in industrial practice. All these methods consist of grinding in rigorously prescribed laboratory conditions in an attempt to simulate commercial conditions. The results achieved generally require corrections based on empirical data.

It is well known that there are three methods widely empolyed in industrial practice, viz. those of Hardgrove, Zeisel, and Bond.

Hardgrove grindability test. In a shaft ball bearing pulverizer according to ASTM D.409, an amount of 50 g of material prepared to 590–1190 microns will be ground. After 60 revolutions the product is sieved on a 200 mesh (74 microns) sieve. The Hardgrove index is calculated according to the empirical formula

$$H = 13 + 6.93\,D \tag{5.1}$$

where D is the passing rate in g. A higher Hardgrove index is equivalent with better grindability, but no proportionality exists between Hardgrove index and energy demand: $H=100$ in no way means a double output when comparing with $H = 50$. The most frequent H values are in the range 40–100. $H = 40$ corresponds to 7.8%, $H = 100$ to 25% passing rate, i.e. the Hardgrove method can characterize only a very coarse grinding. This means that Hardgrove test results can be applied to commercial fine grinding operations only on the basis of experimental data related to similar materials and similar fineness. But it should not be forgotten that the Hardgrove test is a more simple and a quicker method than the two others.

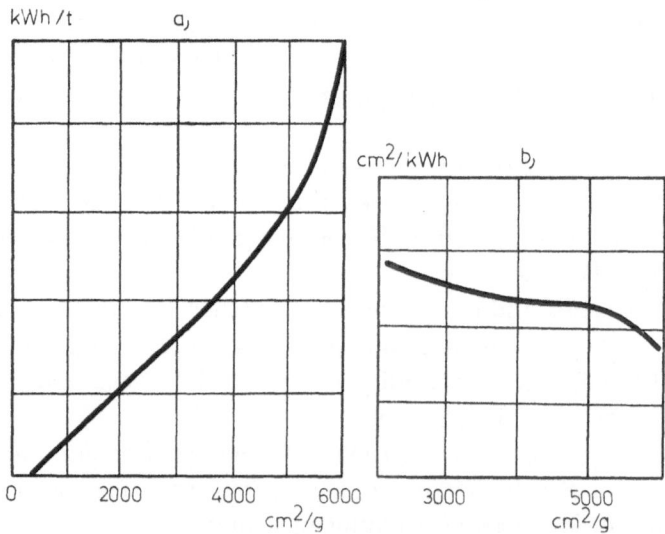

Fig. 5/1. Grindability characteristics of a limestone sample according to Zeisel's method

The Zeisel test (Zeisel 1953, Lehmann and Haese 1955, Bernutat 1961, Ellerbrock 1975). The Hardrove mill was developed originally for coal grinding. Zeisel adapted the equipment to hard materials, and introduced the metric system. But as the most important feature: he did not stop the mill at 60 revolutions but instead he continued the grinding unit attaining the required fineness. The power consumption was determined by torque measurement. The results were presented in a Blaine surface against a power consumption diagram (Fig. 5/1a). The upward bending of the curve represents the deviation from the proportionality according to Rittinger's law. The actual degree of grindability can be characterized by dividing the abscissa by the ordinate (Fig. 5/1b).

Based on widespread investigations carried out with various materials Zeisel was able to state that the commercial operation requires double the amount of energy indicated by the laboratory test. This statement was valid for dry grinding in ball mills of the early fifties.

Recent investigations by Ellerbrock (1975) hinted that the procedure carried out with a semi-technical 0.63×0.63 m laboratory ball mill instead of the Zeisel apparatus gave values in somewhat closer agreement with those of plant mills.

The Bond test (Kannewurf 1957, Bond 1961, Wasmuth 1969). The above-mentioned two methods simulate the open grinding process. Bond, on the contrary, in accordance with the so to say exclusively applied commercial practice in the United States developed his method to simulate closed circuit processes. The test is carried out in a 12 in $\times$ 12 in (305×305 mm) laboratory ball mill and consists of succeeding steps each time replacing the undersize sieved fractions by new original feed of equal quantity. The circuit coefficient is prescribed as 3.5. Steady state is reached when in following steps the same mass G is to be replaced. This replaced mass can be accepted as characteristic of the grindability in real conditions. The empirical formula to calculate the work index is

$$Wi = \frac{16}{G^{0.82}} \sqrt{\frac{h}{100}} \quad \text{kWh/sht} \tag{5.2}$$

h being the mesh size in microns of the test sieve.

The energy demand of the commercial operation can be calculated according to (1.3b) as

$$W = 10Wi \left(\frac{1}{\sqrt{P}} - \frac{1}{\sqrt{F}} \right) \quad \text{kWh/sht} \tag{5.3}$$

where P is the undersized 80% of the product, F that of the feed.

Bond later modified somewhat the above formulae and recommended

$$Wi = 44.5/h^{0.23} G^{0.82} \left(\frac{10}{\sqrt{P}} - \frac{10}{\sqrt{F}} \right) \quad \text{kWh/t.} \tag{5.4}$$

Here P is the value in microns which 80% of the last cycle sieve undersized product passes, F the size which 80% of the new ball mill feed passes.

Formula (5.4) is equivalent to (5/2) in the case of sieve mesh 150 ($h = 105$ microns).

The Bond test is the best simulation of up to date closed circuit grinding processes in ball mills, the decisive role of the mill size remaining, however, beyond consideration. The prescribed 3.5 circuit coefficient is equivalent of a sieve oversize of about 70% in the mill, a very coarse grinding fineness even in closed circuit process, e.g. for cement grinding.

Hardgrove grindabilities can be converted with satisfactory approximation to Bond's work indices using the following empirical formula (Bond 1961)

$$Wi = 435/H^{0.91} \qquad (5.5)$$

As a conclusion it can be stated that no absolute value exists characterizing grindability. The probable energy demand of commercial grinding operations can be calculated by usual laboratory tests only with the application of correction factors based on ample experience. Concerning the grinding of raw cement material the practicability of the three test methods was analysed by Haese et al. (1975).

Grindability and particle size distribution. The author has published in two papers (Beke 1970, 1975) his finding that there exists a qualitative relationship between grindability and the scatter of particle size distribution, or which is the same, the uniformity coefficient n in the RR distribution formula. A smaller scatter or a higher uniformity coefficient is equivalent to a worse grindability requiring higher energy expenditure.

To our knowledge Callcott was the first who noticed this regularity in testing Australian coals (Callcott 1968).

Anselm, in his monograph published in 1950, considered the uniformity coefficient as a material characteristic. In one of his tables the sequence of uniformity coefficients was clearly a grindability sequence without, however, this connection being mentioned. It is, however, a disturbing phenomenon; the uniformity coefficient is not independent of grinding fineness—as was detailed in Chapter 4. In the early phase of grinding the uniformity coefficient has a slight increase, this is first followed by a slight and later by an abrupt decrease signalling agglomeration. Agglomeration is a consequence of energy oversupply: grinding impacts of higher individual energy cause early agglomeration. The reverse of this thesis is valid too: materials with a low uniformity coefficient are inclined to early agglomeration because of their high content of small particles. To eliminate the disturbing effect of agglomeration as well as the predominant role of feed in the early phase of grinding it seems expedient to take into consideration the stage at R(0.09) = 10% as proposed by Anselm.

38

An exception is to be allowed if in this relatively early stage agglomeration occurs so the low uniformity coefficient indicates agglomeration and not good grindability. A typical material manifesting this misleading phenomenon is βC_2S (dicalcium silicate).

According to widespread views grindability is governed by hardness, strength, elasticity and porosity.

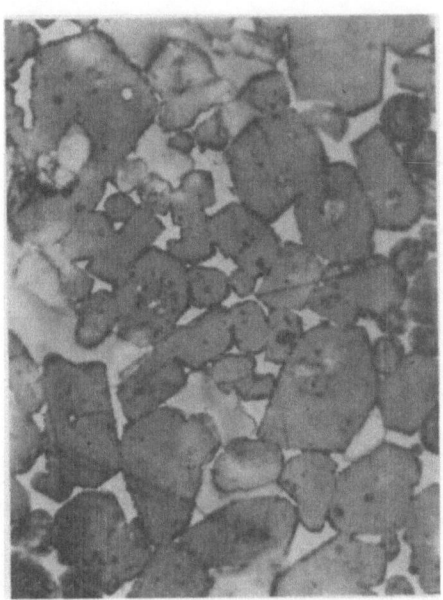

Fig. 5/2. Micropolish of alitic clinker

In the following we hope to demonstrate that the size distribution, or which is the same, the grindability, is affected mainly by the crystal structure which mostly neutralizes the effect of the above properties. In handbooks and test reports we can find data supporting this finding.

If the structure allows the free movement of crystals side by side and the scatter of size is great, such as it is with limestone, the initial grindability is good but there is a tendency to agglomeration.

On the other hand if the crystals are nearly uniform in size but angularity does not permit free movement, crystal breakage is unavoidable in the initial phase. Such is the case with quartzite or alitic cement clinker (Fig. 5/2). Following initial breakages the possibility of free movement increases, grindability improves — as indicated, for example, in formula (4.6) —, agglomeration occurs only in a later phase.

In Anselm's cited work we find for *portland cement clinker* of varying grind-ability at $R(0.09) = 10\%$, n values in the range 0.85–1.13; the generally accepted mean value is unity. In *Handbook of mineral dressing* (Taggart, 1945) we find for American clinkers the Hardgrove indices 31–79, the average being 55. Numerous tests with Hungarian clinkers gave values in the range 41–93, with the mean value of 50.

Uniformity coefficients of *limestone* and *cement raw meal* occur, according to Anselm, between 0.70 and 0.93; Hungarian tests resulted in 0.6 to 0.8.

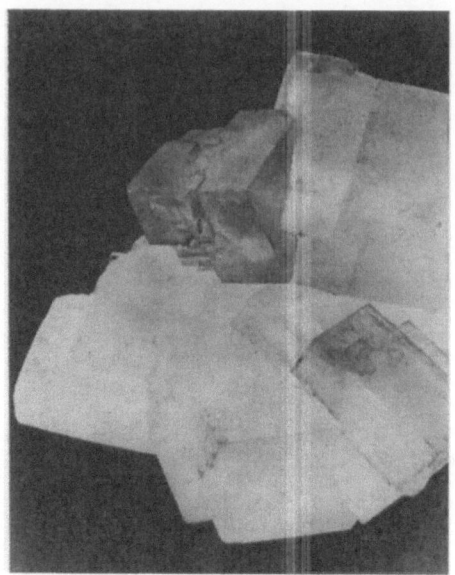

Fig. 5/3. Crystal structure of rock salt

Hardgrove indices in Taggart's handbook are given as 54 to 78, in Hungary these are 68 to 132.

As for *gypsum* which is known for its good grindability, a test exhibited $n = 0.77$ and $H = 114$.

The effect of hardness was investigated by parallel tests with limestone (Mohs hardness 3–3.5) and rock salt (Mohs hardness 2). The hard limestone proved to be more easily grindable than the soft rock salt. For rock salt we obtained $H = 46$ (Taggart $H = 54$), uniformity coefficient about 1. The crystal structure gives the explanation: crystals of rock salt bite into each other (Fig. 5/3), and there is no movement without breakage; on the other hand, limestone crystals are freely moving—at least in the initial phase.

The role of strength and elasticity can be illustrated by the behaviour of basalt and andesite (Hungarian samples). A random test with samples of average quality resulted in

basalt $n = 1.07$ $H = 50$
andesite $n = 0.98$ $H = 64$

According to handbook data, Young's modulus and average compressive strength are, for basalt: $E = 600\,000$ kgf/cm² (60 000 MPa), $\sigma = 2000$ kgf/cm² (200 MPa), for andesite: $E = 250\,000$ kgf/cm² (25 000 MPa), $\sigma = 1500$ kgf/cm² (150 MPa). The characteristic value for elastic deformation work $\sigma^2/2E$ is for basalt 3.3, for andesite 4.5. As a consequence, basalt should be broken more easily than andesite, and in fact it is, e.g. with jaw crushers. But andesite

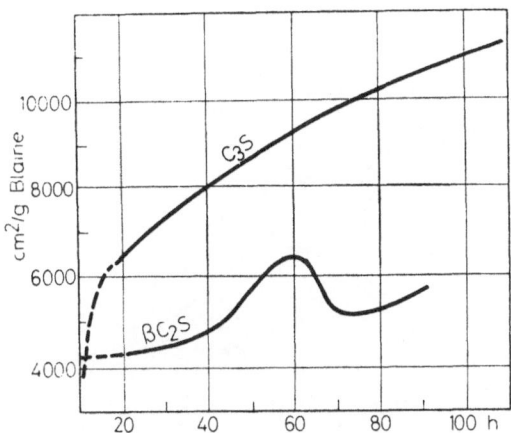

Fig. 5/4. Batch grinding in laboratory ball mill of C_3S and $\beta\,C_2S$. Surface against grinding time

grinding had lower energy expenditure as indicated by n and H values too. Micropolish of basalt bodies showed that the crystals of this material are closely packed, crystals of andesite were less densely packed, could move more easily and are of scattered sizes.

Widespread investigations into the effect of porosity have not furnished unambiguous findings. Summarizing previous research activity, Deckers (1972) stated: grindability in the coarser range improves with increasing porosity, whereas for fine grinding (cement fineness) there is no influence of porosity upon grindability. According to other opinions (e.g. Opoczky and Mrakovics 1976) not the whole pore volume but the size of the individual pores will determine the effect of porosity; higher microporosity (1–25 µm) will facilitate the grinding process.

We shall now discuss the irregular behaviour of βC_2S mentioned above.

In Fig. 5/4 progress of grinding is illustrated together for C_3S and βC_2S. It is a well known fact that C_3S has a far better grindability in spite of its higher uniformity coefficient.

For βC_2S we were able to establish $H = 63$ Hardgrove index and in the laboratory ball mill $n = 0.91$ uniformity coefficient. Both data indicate favourable grindability which is, however, illusory.

41

But the Hardgrove index is, as has been stated above, characteristic only for the initial stage of grinding (in this case, residue on the test sieve is 86%) and the favourable uniformity coefficient demonstrates early agglomeration and not good grindability. To explain this latter statement let us examine Fig. 5/5 representing the crystal structure built of about 40 microns diameter spherically-shaped crystals. Only a very slight displacement is possible, afterwards the crystals must be compressed on all sides, agglomeration is unavoid-

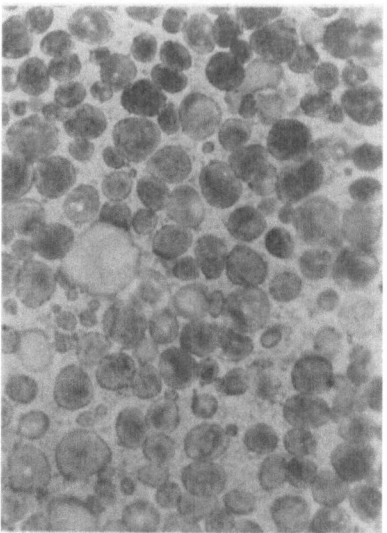

Fig. 5/5. Micropolish of $\beta\,C_2S$

able. The specific surface of the spherical crystals is about 500 cm²/g. Allowing a further 500 cm²/g for the crumbling of the intermediate phase it is clear that the limit of agglomeration-free grinding is at about 1000 cm²/g. In Fig. 5/4 this is indicated as extrapolation by dashed lines.

In Fig. 5/6 we present an interesting picture showing agglomerating particles of βC_2S under the electron microscope. The results of the above experiments and considerations point to the uniformity coefficient as a better characteristic of grindability than hardness or strength; the uniformity coefficient at 10% oversize to 90 microns can be accepted as a material characteristic. However, if at this fineness agglomeration occurs, a small uniformity coefficient can be misleading, in this case it does not indicate good grindability but a tendency to early agglomeration.

It therefore seems justifiable to accept the uniformity coefficient as a quick item of information on grindability. As an explanation, the following factors could be considered:

— materials of good grindability and low uniformity coefficient have a structure with great scatter of crystal sizes,
— difficult grindability requires a long grinding time, and a great impact number which causes the growth of the uniformity coefficient (Fig. 4/2).

The dominating role of the uniformity coefficient can be explained by the fact that no accurate energy dosing is possible, some overdose is unavoidable thus surpassing the effect of strength and hardness.

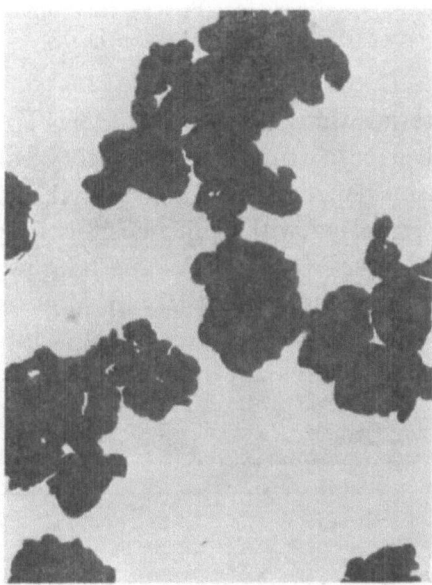

Fig. 5/6. Agglomerate particles under the electronic microscope

The important role of the uniformity coefficient is unambigouus, it gives quick information but to get numerical values to the energy demand of commercial operations the discussed laboratory methods — in spite of their imperfections — are to be preferred.

With regard to the phenomena of aggregation and agglomeration and their connection with grindability Opoczky (1977) differentiates three types of materials

— easily grindable materials tending to aggregation (e.g. limestone),
— materials of difficult grindability tending to aggregation and agglomeration (e.g. cement-clinker, or more explicitly, C_2S),
— materials of difficult grindability not tending to aggregation (e.g. quartz).

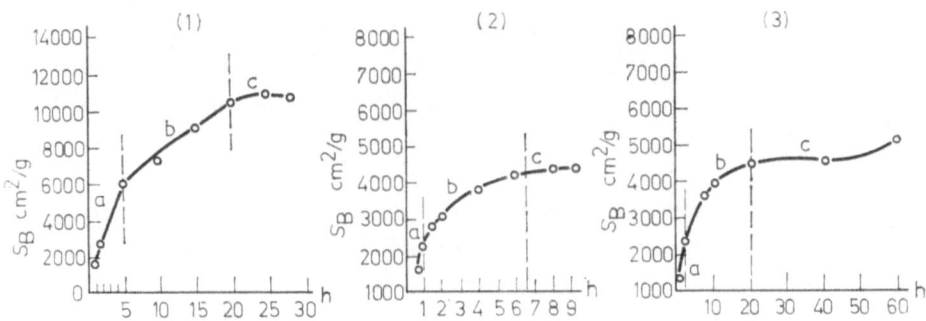

Fig. 5/7. Change in specific surface S_B as a function of grinding time
(1) quartz, (2) cement clinker, (3) C_2S

For these three materials this is demonstrated in Fig. 5/7 by a diagrammatic Blaine surface against grinding time in a laboratory ball mill (Opoczky 1977).

To conclude these considerations, in Table 3 work indices of some materials are listed (Bond 1961) together with their densities. Density has an important influence: lower density gives a greater volume for a ton of material, therefore the same mesh has a larger surface. The listed values are averages related to several American material samples, scatter is not indicated.

Table 3

Average work indices according to Bond (1961)

Material	kWh/t	kg/dm³
Cement clinker	13.49	3.09
Cement raw mix	10.57	2.67
Limestone	11.61	2.69
Quartz	12.77	2.84
Sandstone	11.53	2.68
Silica sand	16.46	2.65
Blast furnace slag	12.16	2.39
Clay	7.10	2.23
Coal	11.37	1.63
Gypsum rock	8.16	2.69

In a later communication, in an attempt to provide a better approximation in commercial circumstances, Rowland (1975) set forth in detail correction factors taking into consideration the effects of dry and wet grinding, open and wet grinding, open and closed circuit, mill shell diameter, oversized feed, very fine grinding, variation of reduction ratio, differences between ball and rod milling.

Symbols in Chapter 5

H	Hardgrove index
D	passing rate, g
G	grindability characteristic, g
Wi	work index, kWh/sht or kWh/t
h	sieve mesh, microns
P	characteristic product size, microns
F	characteristic feed size, microns
n	uniformity coefficient.

CHAPTER 6

GRINDING OF MATERIAL MIXTURES

Manufacturing processes often require ground products of mixtures of materials (cement raw meal = limestone + marl, blast furnace cement = clinker + blast furnace slag, pulverized coal firing often uses a mixture of various sorts of coal). The raw materials themselves, of course, consist of naturally mixed minerals too.

To grind mixtures of materials there are two possibilities: grinding separately followed by subsequent blending, or common grinding. Blending of pulvers is carried out by pneumatic methods with considerable energy expenditure. Common grinding supplies a more homogeneous product. But in common grinding the components exercise an effect on each other, the one component can help or hinder the comminution of the other; sometimes this effect varies with the mixture rate and/or particle size. This means that the blending is complete only for the whole mass, for individual size fractions there will occur a selective process: in smaller sized fractions the one, in larger sized fractions the other component will be represented in a higher proportion. This phenomenon can reach such a measure that the components can be separated by classification. Separation of natural mineral mixtures by this method is called selective grinding. Selectivity is of course often unwanted, in this case separate grinding is the only solution.

Particle size distribution of *separate ground mixtures* of materials a and b is described by the formula

$$R = ce^{-\left(\frac{x}{x_a}\right)^{n_a}} + (1 - c)\, e^{-\left(\frac{x}{x_b}\right)^{n_b}} \tag{6.1}$$

where c is the mixing rate, which is obviously impossible to transform to the form (3.2). It means that size distribution of milling product mixtures cannot be described by the RR (or GS, etc.) formula or represented by straight lines in the RR chart. The particle size distribution of blended milling products is represented in Fig. 6/1 in the RR chart, left side $n_a = n_b$, right side $n_a \neq n_b$ (Reményi, 1966).

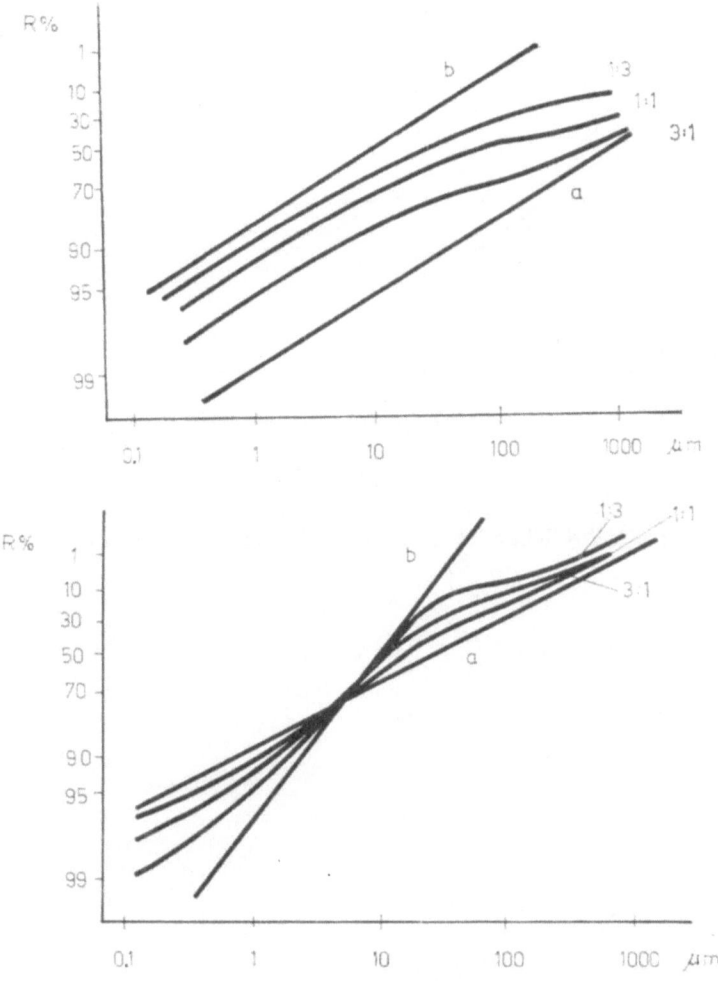

Fig. 6/1. Particle size distribution of separate ground mixture in RR chart. Left side: $n_a = n_b$; right side: $n_a \neq n_b$ (Reményi 1966)

In *common grinding*, as has been said above, the mutual effect is not negligible, it can promote or hinder the grinding process. In Fig. 6/2 four possibilities are demonstrated in presenting Hardgrove indices against mixing rate: line 1 represents the case of no mutual influence, line 2 presents the promoting, line 3 the hindering effect, in case 4 the effect varies as a function of mixing rate.

Tanaka (1962) distinguishes five types of mutual effect. In Fig. 6/3 the specific surface is presented against grinding time, A_0 and B_0 are related to separate grinding, A and B to common grinding, C to the mixture.

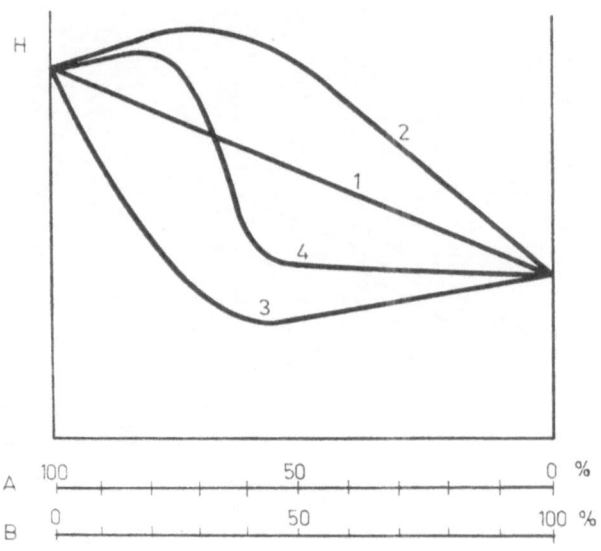

Fig. 6/2. Possibilities of mutual effect in common grinding

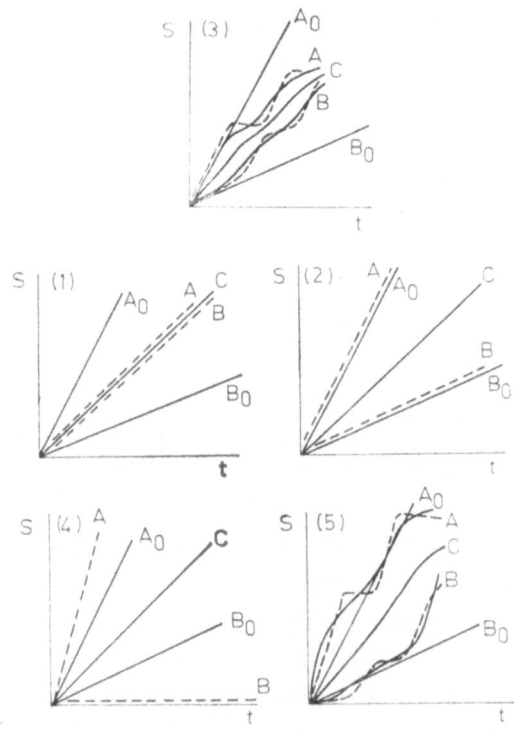

Fig. 6/3. Five possible models of grinding binary mixtures according to Tanaka. A_0 and B_0 separate, A and B common grinding, C surface of the mixture (1962)

In case 1 the mutual influence results in identical grindability of the components; in case 2 there is no mutual influence at all. In case 3 with change of grinding fineness the components will attain alternatively the dominant role. Case 4 represents the perfect selectivity, component B remains unaffected; finally, in case 5 the selectivity is influenced jointly by the grindability and grinding fineness.

According to investigations by Reményi (1966), in common grinding the particle size distribution of the components follows the same regularities

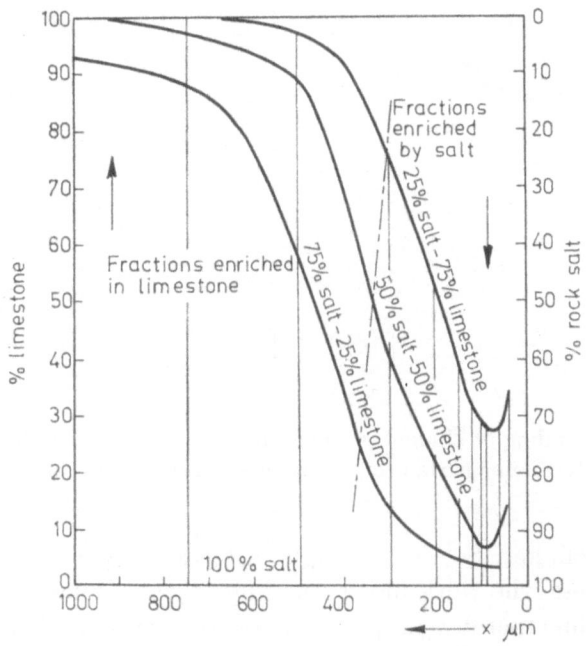

Fig. 6/4. Common grinding of limestone and rock salt. Composition of size fractions (Reményi 1966)

(e.g. the RR formula) as in separate grinding, but with differing constants. As a consequence, the particle size distribution of the mixture cannot be demonstrated by straight lines in the same chart (as seen in Fig. 6/1).

Because of the common influence of particle size an enrichment of one component takes place. Figure 6/4 demonstrates the composition of common ground rock salt and limestone mixture against particle size (Reményi, 1966). The grinding was executed in a Hardgrove mill with 60 revolutions. (It is interesting to note that: the samples in this test had grindability indexes not to be met in the manuals, salt had a better grindability ($H = 65$) than limestone ($H = 46$). But this does not influence the important finding concerning the particle-size dependent enrichment in one component.)

The effect of different ratios of fine particles is demonstrated in Fig. 6/5. Clinker and blast furnace slag were commonly ground according to the prescriptions of the Bond test. On the right side the work indexes, on the left side the Blaine surface of undersize are represented as a function of the mixing ratio. It is somewhat surprising that the side of worse grindability (higher work index) supplies the larger surfaces. It can be explained by the fact that the surface of undersize (about 30% of the mixture) is determined by the

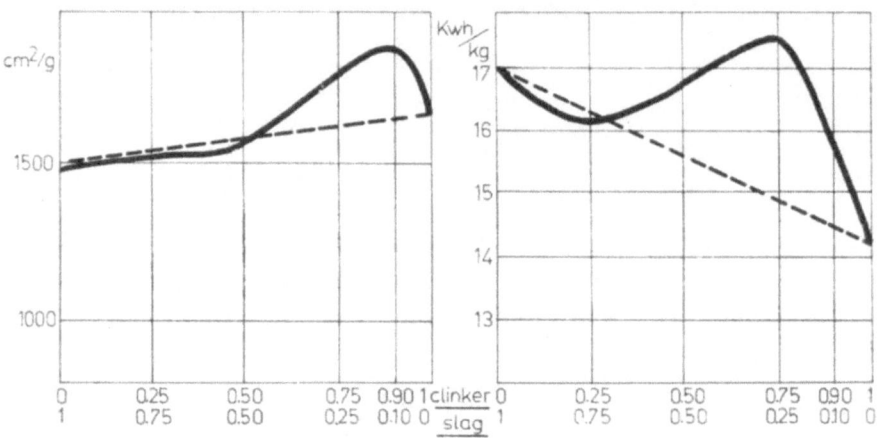

Fig. 6/5. Common grinding of clinker and blast furnace slag. Right: Bond's work index; left: Blaine surface of undersize against mixing ratio

better grindable clinker. So, for example, a small proportion of slag impairs the grinding process, the work index gets higher, the clinker will have a prolonged grinding time thereby supplying more fine particles and a larger surface. On the other hand, clinker added to slag improves grindability, smaller energy expenditure results in smaller surface.

Our considerations result in the finding that because of the mutual influence of mixtures of materials no grindability index can be determined; moreover, the concept of grindability cannot be interpreted at all.

Symbols in Chapter 6

    R sieve residue
    c mixing ratio
    n uniformity coefficient
    x particle size, microns
    x size module, microns
    H Hardgrove index

CHAPTER 7

GRINDING AIDS AND MECHANOCHEMISTRY

In the course of grinding—especially in its fine phase connected with the complicated energy transformations—structural and chemical changes take place. In (c) in Chapter 1 (definition (c)) it was hinted that there is a limit of the fracturing effect and in Chapter 4 the phenomenon of agglomeration was dealt with in detail. With these factors in mind, grinding and especially its fine phase should be considered not just as a simple mechanical process but one that is also partially a physicochemical process.

The phenomena of mechanochemistry were dealt with in detail by Peters (1962), a good summary of related literature is to be found (in Hungarian) in a paper by Menyhárt (1972).

Lattice forces are in a state of equilibrium inside the body but on the surface they become unsaturated, the surface energy or surface tension is not bound and can become effective in the boundary layer of the solid body and the neighbouring gaseous substance.

Let us quote the finding of Götte (1952) one of the originators of this part of grinding theory:

"Surface tension is due to the fact that while ions, atoms and molecules combine into a compact lattice inside the body the pattern is less saturated at the outer surface giving rise to increased mechanical strength in that region and inducing forces of considerable magnitude in the boundary belt of solid and gaseous substances. The production of new surfaces charged with such forces will require large amounts of surplus energy. If, however, free surface energies are affected by the addition of adsorptive liquid molecules, grindability can be improved considerably." (See e.g. Fig. 4/5.)

In practice, these boundary forces can be manifested as agglomerations or as the increase of surface reactivity. This latter effect is quoted as mechanical activation and is executed mostly by vibration milling or more effectively by jet.

Agglomeration can be retarded, as mentioned above, considerably by adding small quantities of tensioactive agents such as amines, glycols, technical alcohols, etc. Water vapour also has a slight disaggregating effect which tends to cause separation of agglomerates and this is widely applied in cement grinding; apart from this, water vapour improves the effectivity of electrostatic precipitators.

Rehbinder (1944) was the first to discover the effectivity of tensioactive materials. He mixed electrolytes in the ratio of 0.01–0.1% to the rinsing water of drilling devices and was able thereby to obtain a 20−60% increase in boring velocity. According to his explanation comminution occurs on the repeated influence of mechanical impacts but in the time between two impacts the fissures become welded because of the surface tension of both sides. This phenomenon can be prevented by the presence of tensioactive agents. This is the so called and often disputed "Rehbinder effect" or "strength decrease by adsorption".

Several scientists deny the existence of the Rehbinder effect and explain the improvement of the grinding process mainly by the increased mobility of particles because of reduced inner friction (Seebach 1969).

With factory practice in mind, it can be stated that grinding aids are advantageous in the coarse stage of grinding too (e.g. in the first chamber of multichamber mills), where the surface has reached only a moderate value, so the surface properties cannot here explain the improved grindability.

Ocepek and Eberl (1975) summarized this disputed problem and completed the picture by making a comparison between the velocity of the fracture front in the solid particle and the spreading velocity of the grinding aid.

Without intending to settle the dispute, an explanation of the effectivity of grinding aids can be given by the following phenomena:

(a) Rehbinder effect. The increase of grinding velocity in the coarse phase can be explained as a consequence of this effect.

(b) Hindering or even eliminating the adhering of ground particles on lining and grinding bodies.

(c) Retarding aggregation and therefore the subsequent agglomeration of previously ground particles.

(d) Change in fluidity properties, decrease of internal friction in the ground mass, attaining dominant role in the fine phase as dealt with in several papers, e.g. Seebach (1969), and Scheibe et al. (1975).

Because of their high price, the utilization of grinding aids is mainly a financial matter. In cement grinding (according to Hungarian circumstances) surpassing 3000 cm^2/g Blaine surface amine-based grinding aids proved to be economical. Over about 4000 cm^2/g the application of grinding aids is unavoidable.

Plant experiences in the GDR (Richter et al. 1974) resulted in a 15–20% output increase at unchanged grinding fineness; the grinding aid was 350–

500 g/t octane diol, dosing took place by dripping into the mill inlet avoiding the more complicated pulverizing procedure. A comparison between open and closed circuit processes revealed that with an open circuit agglomeration is retarded, the mill length seems to be—and function as if—enlarged; some increase in output can be achieved but the main purpose is to ensure the required high degree of fineness. In a closed circuit the fineness is governed by the classifier; by retarding agglomeration the recirculation decreases, more raw material can be fed into the mill, and the output increases. But because the tailings (coarse fractions) take away a part of the grinding aid, a surplus dose is needed. One observation should be mentioned here: dripping cannot be effective near the inlet side, it is effective only following vaporization.

Some other observations are that the milling product flows more easily, the slope of conveyor band needs to be reduced, the retention time in the mill decreases and in consequence the grinding media charge has to be rectified (smaller grinding bodies are required); the collection of dust is a thorny and complex problem; and because the electrical resistance of the powder increases, water pulverizing in needed.

A particularly significant advantage is that because of easy flow, the efficiency of the classifier improves.

Details of Japanese experiences with grinding aids have been published in several papers (e.g. Jimbo 1972, 1973, and Iwabuchi 1968); the effects of gaseous and solid additives were compared. In cement grinding the increase in the coefficient of uniformity was established; this increase favourably affects the hardening process (see Chapter 19).

A paper by the Soviet scientists Nudel and Krikhtin (1976), analysed the effect of surface active agents on the grinding of cement raw materials. The authors stated that the decrease in strength by adsorption according to the Rehbinder principle takes place thus improving the effectivity of the grinding process and increasing the mass of fine particles. The determining properties of the grinding aids are the adsorption activity which characterizes the capacity to reduce the surface energy of the solid body and the surface occupied by the grinding aid molecules which determines the penetration into the defect locations of the crystal lattice.

It is appropriate here to add a few words about mechanical activation, or intensification of chemical reactivity. As has been said above the principal method of mechanical activation is vibration or jet milling. The explanation is that the energy is furnished by light grinding bodies or feed particles describing short trajectories so the overdose of individual energy impacts causing agglomeration is avoided, the surface therefore remains unsaturated.

Effective activation can be attained by high speed pin mills or cutting mills too. The explanation is self-evident: no contact takes place between the pulverized particles of high energy concentration.

According to Juhász (1974) three main groups of mechanical activation can be distinguished.

Among the first group, reactions can be ranged where crystal transformations take place without, however, any change in the gross chemical composition. These processes are characterized by modification from the stable to the unstable state, amorphization and energy level increase takes place; this is the case of real activation.

The activation of quartz was dealt with in several papers by Schrader (e.g. 1968), that of lime by Schrader and Hoffmann (1970).

Rehbinder and Chodakov (1962) demonstrated that activity is affected mostly by the degree of amorphization and not by the attained dispersity.

As an example, Fig. 7/1 presents X-ray diffractograms of clinker; in item *a* following 1 hour, in item *b* 90 hours grinding time in the laboratory ball mill. The Blaine surfaces were, respectively, 1800 and 10 000 cm²/g, the flattening of peaks demonstrates amorphization. To attain the high grinding fineness, the grinding aid triethanolamine was applied (Beke and Opoczky, 1967).

Another example is given in Fig. 7/2, which illustrates an experiment by Juhász (1972). It shows the DTA diagram of dolomite grinding in a vibration mill, item *a* before grinding, item *b* after a grinding time of 32 hours. It is interesting that the translation of both "valleys" are connected with de-carbonization, the three exothermic peaks demonstrate an energy level in-crease.

This kind of activation was studied with different types of grinding equipment in the FIA Institute (GDR) (Bernhard and Heegn, 1975). It was shown

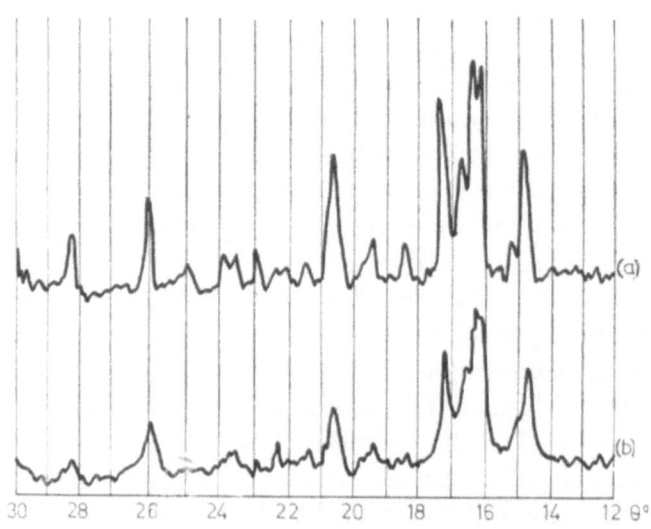

Fig. 7/1. X ray diffractogram of ground clinker:
(a) grinding time 1 h, specific surface 1800 cm²/g; (b) grinding time 90 h, specific surface 10 000 cm²/g

by some examples that the individual mill parameters as well as the different mill types lead to a characteristic relationship between fracture phenomena and changes of the crystal lattice of solids.

These principles are utilized in the production of a new type of building material the "silicalcite" invented by Hint some twenty years ago (Hint 1976).

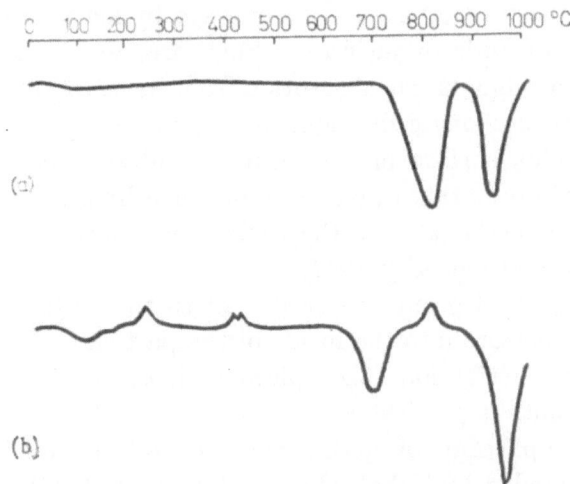

Fig. 7/2. Dolomite grinding in vibration mill, DTA curves
(a) initial stage, (b) following 32 h grinding (Juhász 1974)

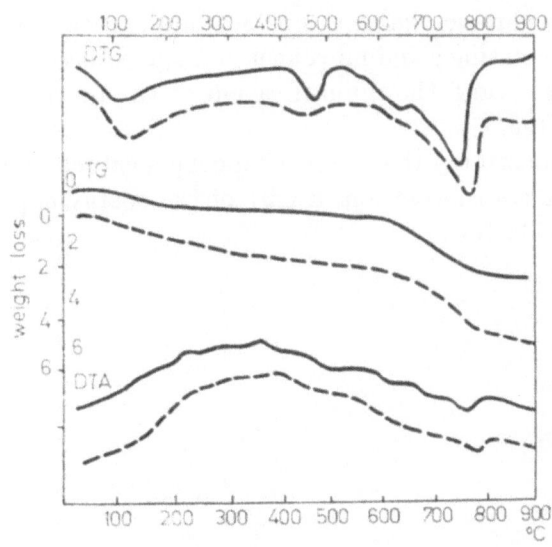

Fig. 7/3. Cement grinding in laboratory ball mill. Derivatogram following 1 h (full lines) and 70 h (dotted lines)

This is a steam cured product of ground sand and lime. It was stated that if sand was ground in a cage mill instead of in a ball mill, the strength of the product was some 2.5 times higher—obviously due to mechanical activation. The result can be explained by the higher energy level of the individual impacts without, however, allowing the contact of the cloudwise dispersed particles, therefore the high energy level does not give rise to agglomeration.

The second group of reactions is characterized by material transport on the particle surfaces, inwards or outwards. Such are, for example, solid phase reactions during grinding or gas absorption from the environment. In Fig. 7/3, in a derivatogram, carbonate development during long term grinding is demonstrated: the high surface particle fractions absorb water vapour in the milling space, hydrolysis takes place forming lime hydrate, this latter transforms into calcium carbonate by CO_2 absorption, again from the milling atmosphere (Beke and Opoczky 1967).

Processes in the third group are restricted to the surface layer; diffusion velocity cannot penetrate into the inside of the particles.

Effective surface activation takes place in loose nodulized aggregates in consequence of pointlike particle contacts.

An interesting application of mechanochemical effects on zinc oxide powder crystals was expounded by Takahashi and Tsutsumi (1967). When subjected to mechanical treatment the crystallic size and lattice distortion measured by X-ray diffraction method, the BET specific surface, the density and the zeta-potential showed monotonous changes in relation to treatment time.

An analogous procedure was investigated and patented in Hungary to produce zinc, lead and chrome pigments without liquid phase components. It was stated that with increasing grinding velocity and grinding fineness an increased frequency of contacts and the intensification of the process can be achieved (Menyhárt et al. 1965).

It has been demonstrated that all mechanochemical reactions are connected with crystal lattice transformation: a part of the energy supply increases the energy level of the lattice.

CHAPTER 8

ENERGY DEMAND AND EFFICIENCY

It is a common fault of the earlier discussed Rittinger, Kick, and Bond comminution theories that they characterize the mass of particles with a single size (e.g. x_{80}) and the other statistical characteristic, i.e. the scatter, is neglected.

An interesting coincidence is that in one and the same year (1957), three papers (Charles, Holmes, and Svensson and Murkes) put forward a new formula with the aim of eliminating this deficiency. This formula can be written with our symbols as

$$W = C \left(\frac{1}{x_2^{n_2}} - \frac{1}{x_1^{n_1}} \right) \tag{8.1}$$

where W is the energy demand, x_1 and n_1 the size module and the uniformity coefficient of the feed, respectively, and x_2 and n_2 those of the product, C is a constant. If $n_1 = n_2 = 1$ we arrive at the Rittinger formula; if $n_1 = n_2 = 0.5$, at Bond; if $n_1 = n_2 = 0$ as the limit value, at the Kick formula.

If we consider the change of n according to formula (4.4) and Fig. 4/2, in the case of agglomeration n_2 can be lower than n_1, and (8.1) can result in a negative value. Formula (8.1) must therefore be modified. After due consideration the following empirical formula was developed

$$W = C \left(\frac{v_2}{x_2} - \frac{v_1}{x_1} \right) \tag{8.2}$$

where v is the variation coefficient of the distribution (quotient of standard deviation and mean value)

$$v = \sqrt{\frac{2}{n}! - 2 \left(\frac{1}{n}! \right) + 1} \tag{8.3}$$

n the uniformity coefficient in the RR distribution (Beke 1962).

It is important to note that in the phase of fine grinding preceding agglomeration the Rittinger formula is well applicable. However, it should not be forgotten that the calculated value does not represent the real comminution work, rather that of the energy losses. This is acceptable: the friction work transformed to heat is determined by the surface of the milling product.

Where $n_1 = n_2$, our formula simplifies to (1.1) i.e. that of Rittinger.

In the following we shall discuss the, so to say, unsettled question of grinding efficiency.

Generally speaking, efficiency is a fraction with the work input in the denominator, its utilized part in the numerator; or the difference of the whole and the wasted energy.

In the science and practice of energy transforming machinery (e.g. steam engine or turbine, electric motor or transformer, pump, compressor, etc.) the calculation is unequivocal: wasted energy manifests itself as heat.

Energy transformation is not the object of comminution operations (just as with, for example, metal working, briquet pressing, loom or other textile machines), so the practical or even the theoretical justification of the introduction of the concept of grinding efficiency is disputable. Nevertheless, it was introduced by Smekal (1937) in extending Rittinger's principle. According to Rittinger the result of comminution is the new surface, according to Smekal the energetic result is the new surface energy. This interpretation furnishes efficiency values of some tenths of one per cent. For example, for quartz the specific surface energy is about 10^3 erg/cm^2, the energy consumption in a commercial mill in order to attain the fineness of about 7000 cm^2/g BET is about 30 kWh/t, so the calculated efficiency is 0.7%.

This result was first disputed by Schellinger (1952). With reference to the above machinery he investigated the wasted heat. The result of his well prepared experiment is apt to arouse speculation: he determined the introduced energy by torque measurements, and placed the laboratory ball mill in a calorimeter. Filling up the mill with grinding media only (without any material to be ground), he was able to recover the whole amount of the introduced energy in the form of heat. But when grinding brittle materials in the mill he discovered a 10 to 20% shortage, that is the "efficiency" was even that much.

Similar shortages in the energy balance of laboratory and commercial grinding operation were established in investigations of several research workers.

This problem was discussed in detail in the Second Symposium on Size Reduction (Amsterdam 1966). Papers were presented by Rose (1967), Hiorns (1967), and Hukki and Reddy (1967) which led Hukki and Reddy to formally distinguishing two different efficiencies: one related to the surface energy, the other to heat development.

So the deficiency in heat development hints at the increase in energy content of the material in consequence of energy input by comminution. This energy increase is due t ⸱ mechanochemical processes (mechanical activation, see Chapter 7), to changes in the energy level of the crystal lattice.

It is of interest to note that if we accept Smekal's concept of the analogous operation of metal cutting where similarly new surfaces are created by mechanical energy expenditure, the industrial "efficiency" demonstrates itself as being two orders of magnitude lower than that of brittle material grinding — something never mentioned in the literature. In a pressing operation we should obtain a negative value and an efficiency for, say a loom, cannot be interpreted at all.

These problems have been dealt with in detail in the author's previous papers (Beke 1973, 1976).

Quite another approach was presented by Straimand (1975). Rejecting the calculation related to surface energy which is unlikely to assist in the advancement of mill design he proposes that in commercial mills the energy usage be related to the energy required to give the same degree of comminution when each particle is separately stressed until it breaks (i.e. the relation of single particle breakage to the collective process). This calculation yields efficiencies in the range of 1% for fluid energy mills to over 80% for roller crushers and 7 to 13% for the most important ball and roller mills.

To conclude our considerations: the calculation of grinding efficiency based on surface energy is, in the viewpoint of industrial practice, of no consequence. There are different definitions leading to very different efficiency values and none of them characterizes the practicability of the grinding process.

Symbols in Chapter 8

W work or energy demand, J

C constant

x particle size, microns

n uniformity coefficient

v variation coefficient

CHAPTER 9

CLASSIFICATION AND ITS INDEX NUMBERS

The grinding operation is often connected with particle size classification (e.g. in closed circuit grinding). It seems justifiable therefore to be acquainted with some index numbers characterizing the effectivity of classification.

In the course of classification the comminution product will be separated into two parts, viz. inferior and superior, related to a prescribed particle size. A perfect separation is of course impossible and this is of special importance for air separators of closed circuit operations.

The following short survey of the qualitative characterization and its index numbers is based upon three papers which include the name of Pethő (Pethő 1971, Pethő and Tompos 1974, Pethő and Smirnow 1976).

For a supposed ideal process the distribution of separated fine and coarse products of the by $F(x)$ distribution function characterized feed are represented in Fig. 9/1a and b. In Fig 9/1a, the quantities $F_f(x)$ and $F_c(x)$ are related to the whole mass of feed whereas in 9/1b to their own mass; $F_{fn}(x)$ and $F_{cn}(x)$ are the normalized distribution functions of the separated products.

Distribution $F_f(x)$ and $F_c(x)$ and normalized distribution $F_{fn}(x)$ and $F_{cn}(x)$ of actual separation products are presented in Fig. 9/2a and b, their $f(x)$ frequency functions in Fig. 9/3.

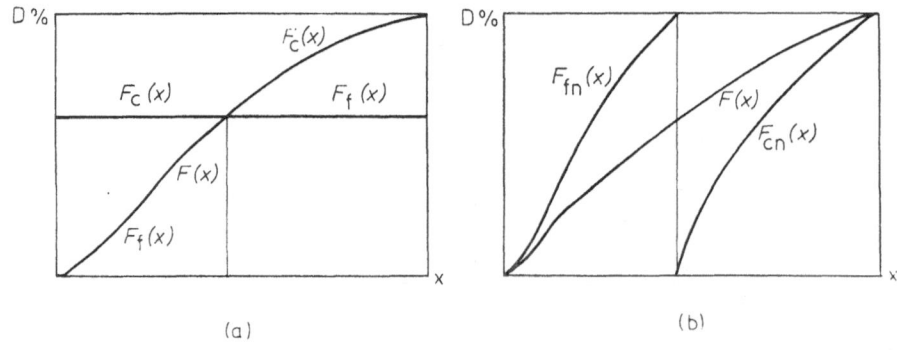

(a)

(b)

Fig. 9/1. Ideal separation process. Particle size distribution (a) and normalized distribution (b) of the feed and separated product

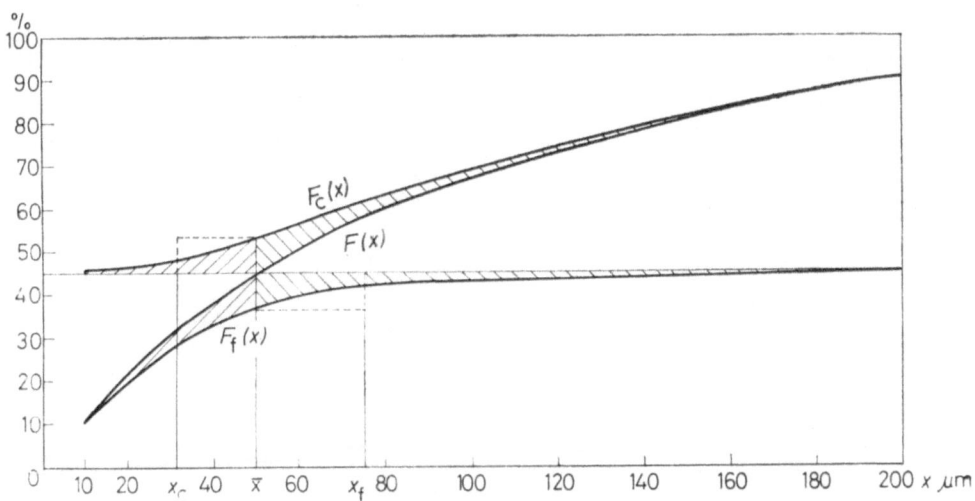

Fig. 9/2a. Particle size distribution in real separation process

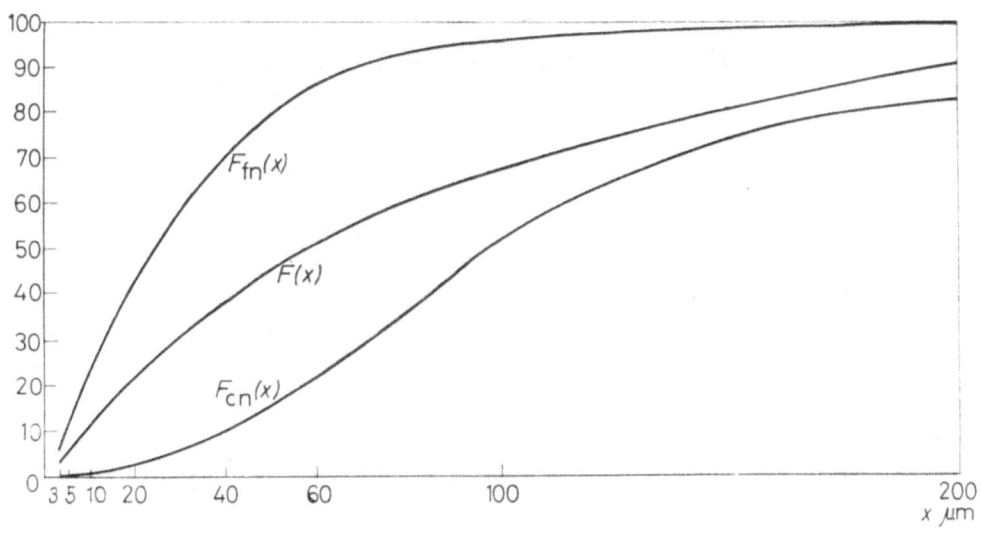

Fig. 9/2b. For caption, see Fig. 9/2a

Representation according to Fig. 9/2a was first proposed by Heidenreich (1954). On the ordinate scale we find the yield of fine class; intersection of its horizontal with $F(x)$ results in $\bar{x}$ the so called equalizing parameter, the natural characteristic particle size of separation quoted by Rumpf (1965) as the analytical size limit.

In these figures ordinate intersections of the distribution function and areas in the frequency function represent mass quotae of the feed and its separated products.

61

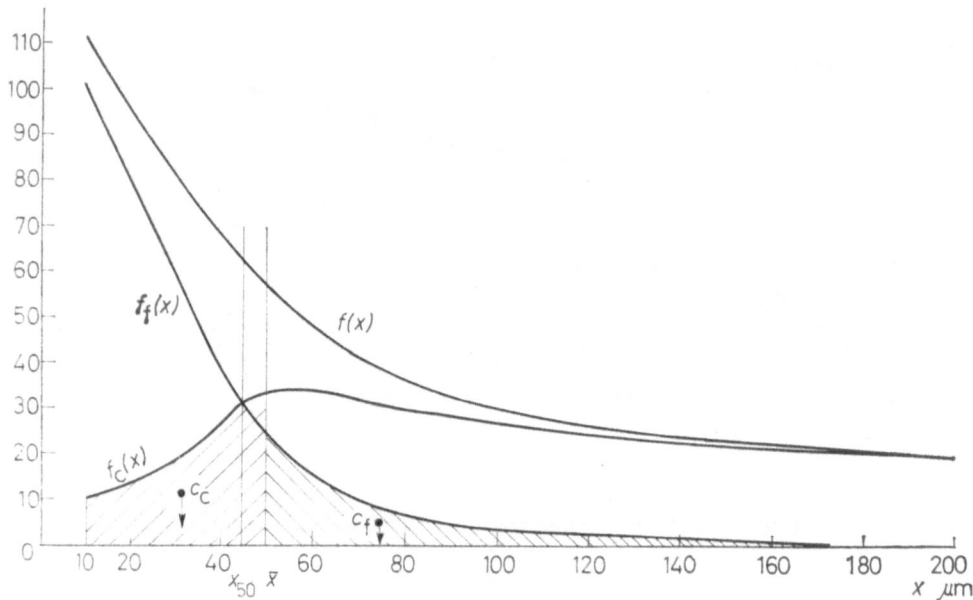

Fig. 9/3. Frequency functions of separation

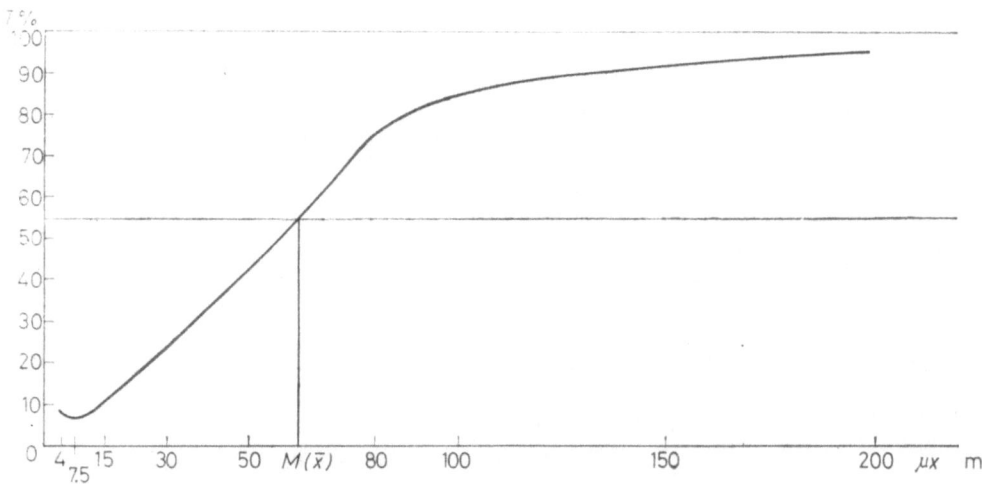

Fig. 9/4. Tromp curve corresponding to Figs. 9/2 and 9/3

In Fig. 9/2a the hatched areas are quoted as the Heidenreich error areas. In this representation it can be read as the respective ratios getting into the fine and into the coarse class. Accordingly the mutually equal ordinate sections between $F_c(x)$ and $F(x)$, and between $F(x)$ and $F_f(x)$ related to the $\bar{x}$ equalizing parameter represent, respectively, the proportion of coarse in the fine and the proportion of fine in the coarse class.

If we transform the Heidenreich areas into rectangles of equal area (Fig. 9/2a), the abscissae x_c and x_f indicate the average quality of the defective parts of the separation products. These same abscissa values in the frequency diagram (derivative of distribution Fig. 9/3) are those of the mass centres of the hatched defective areas.

The intersection of $f_f(x)$ and $f_c(x)$ frequency curves point to the x_{50} median. This particle size (quoted by Rumpf as the preparative size limit) has the same probability of being in the fine or in the coarse class.

To judge the separating process, the curves introduced (intended originally for coal separation by density) by Tromp (1937) are generally applied. These curves point out for every particle size the probability of getting into the fine or into the coarse class. In operations connected with grinding the ordinate $T\%$ usually indicates the probability of getting into the coarse class. The Tromp diagram of perfect separation (see, e.g. Fig. 9/1) is a stepformed line, elevating at the separating size from 0 to 100%.

In the knowledge of the distribution and frequency curves of separation the Tromp curve can be plotted. According to definition

$$T(x) = \frac{f_c(x)}{f(x)} \tag{9.1}$$

or starting from discrete measurement values

$$T = \frac{\Delta F_c(x)}{\Delta F(x)} \tag{9.1a}$$

The Tromp curve corresponding to the preceding figures is presented in Fig. 9/4.

Figure 9/5 represents Tromp curves of a separator of the firm SKET ZAB (Dessau, GDR). The separator is connected to a closed circuit cement mill and data for different loads are given. The deteriorating effect of increasing load is conspicuous; in the range of fine particle sizes the curve bends upwards.

To characterize the Tromp curve, the following index numbers are to be found in the literature:

— x_{50} median, identical probability of getting into the fine or into the coarse class
— "Terra number", calculated from x_{75} and x_{25} quartiles: $E = (x_{75} - x_{25})/2$
— imperfection: $I = (x_{75} - x_{25})/x_{50}$
— $\varkappa = x_{75}/x_{25}$: steepness
— x_m mode, the abscissa of the turning point.

These values, however, are arbitrary, and of no consequence for the quality of separation; what is more, in the case of high values in the range of fine particles they are not even interpretable.

According to Pethő and Tompos (1974) the Tromp curve can be interpreted as the frequency function of one separated product of a uniformly distributed feed. In this manner all the characteristics of an actual separating process (e.g. equalizing parameter, error areas, quality index numbers) can be interpreted and calculated as shown above; full information is given without even needing to plot the curve, and there is also the suitability for computer programming.

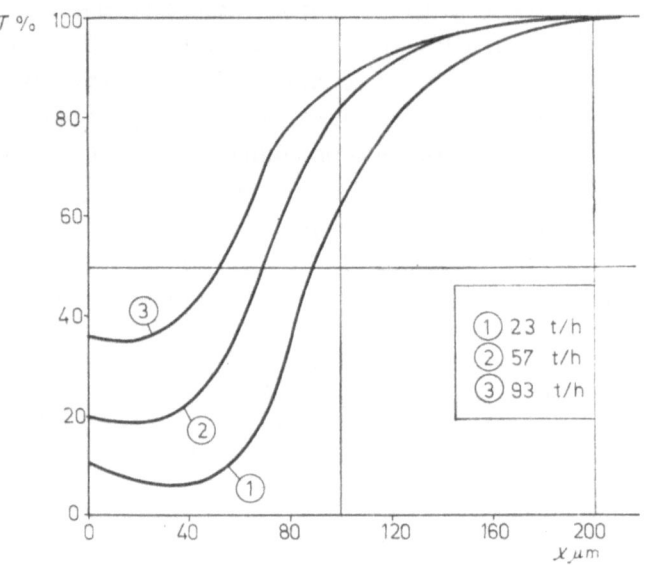

Fig. 9/5. Tromp curves of a separator with different loads (SKET ZAB Dessau)

The most important characteristic will again be the equalizing parameter, the formula of its expected value starting from discrete measurement results in

$$M(\bar{x}) = \Sigma(1 - T_i)\Delta x_i. \tag{9.2}$$

However, the application of this formula in connection with fine grinding encounters difficulties: 100% separation does not occur in the tested range, the whole separation zone cannot be taken into consideration. Here, the equalizing parameter is determined graphically by the horizontal of the proportion of the coarse class.

Because of the predominant significance of the fine particles the use of a logarithmic abscissa scale is desired, accomodating to the RR presentation of particle size distribution. This representation and the related method of calculation is dealt with in the following chapter.

Symbols in Chapter 9

x	particle size, mm or microns
$F(x)$	particle size distribution function
$f(x)$	frequency function
index f	denotes the fine class
index c	denotes the coarse class
index n	denotes normalized function
$\bar{x}$	equalizing size, mm or microns
x_{50}	median, mm or microns
$T(x)$	Tromp's separation function
M	expected value

CHAPTER 10

CLOSED CIRCUIT GRINDING

If we consider a tube mill with its length equal to 5 to 6 times the diameter, it can be stated that at about a quarter of the length, half of the feed has reached the required fineness; in further grinding this mass will be overground or even agglomerated. The remedy has long been known: the grinding process must be interrupted, the fine portion separated, and the coarse portion recycled to the mill. This is the principle of closed circuit grinding.

The layout of a closed circuit grinding apparatus is shown in Fig. 10/1. Feed F in t/h is passed through the mill and then dispatched to a classifier which recycles that fraction G in t/h larger than h micron size to the mill. The throughput D in t/h passing through the mill, the elevator and the classifier is the sum of F and G. In the steady state, fine fraction P of the classifier is equal to the feed, so

$$D = F + G = P + G. \tag{10.1}$$

An essential characteristic to describe closed circuit grinding is the circuit coefficient U, the quotient of throughput and output

$$U = \frac{D}{P}. \tag{10.2}$$

If by random sieve analysis of the three material flows we obtain the oversizes p, d and g respectively, then in view of input–output balance, for the individual size fractions we can write

$$Dd = Gg + Pp \tag{10.3}$$

and in consequence of (10.1) and (10.3)

$$U = \frac{D}{P} = \frac{g - p}{g - d} \tag{10.4}$$

implying that only one of the three flows must be measured quantitatively and the other two assessed by sieve analysis. Since (10.4) is valid irrespective of mesh size the constancy of the circuit coefficient will permit the checking and correction of particle size distribution tests or even their extrapolation.

(It is thought to be necessary to note that these and the following calculations are essentially identical to the calculations in Chapter 9 – though adapted to the practice of fine grinding installations. Here we calculate with sieve over-sizes whereas in Chapter 9 undersizes were taken into consideration – thus, the

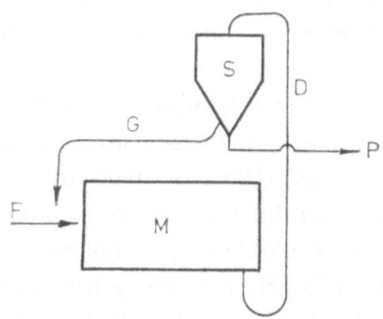

Fig. 10/1. Layout of closed circuit grinding
M mill, *S* separator, *F,D,G,P* material flow t/h

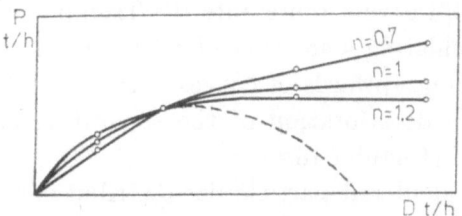

Fig. 10/2. Output against throughput, perfect separation presupposed: real separation indicated by dotted line

use of the different symbols. Circuit coefficient U is equal to the reciprocal of yield of the fine class.)

As a first approach let us suppose a perfect separation at particle size h with sieve residue R. Then for charge P t/h we get the fine class

$$P_f = P(1 - R). \tag{10.5}$$

A glance at Fig. 10/1 is enough to convince us that the same relation exists between output and throughput. If we take (4.9) into account, then the following formula is valid

$$P = D\left[1 - e^{-\left(\frac{ch}{D}\right)^n}\right]. \tag{10.6}$$

Analysis of (10.6) demonstrates the decisive role of the uniformity coefficient (Fig. 10/2).

If

n > 1 the function has a maximum value
n = 1 the function approaches a limit value
n < 1 the function increases monotonically

(This is true only in the ideal case of perfect separation. In real circumstances — in accordance with the characteristics of separation, see Chapter 18 — the curve will always bend downwards and give an extreme value as indicated by the dotted line in Fig. 10/2.)

A low uniformity coefficient has therefore a favourable influence on the quantitative conditions of the closed circuit. In view of this, it is no wonder that for cement raw meal ($n = 0.6$–0.8), the closed circuit operation was introduced some twenty years earlier than for cement grinding ($n = 1$).

In actual practice perfect separation does not take place. Therefore, depending on their size, particles will get, with certain probability, into the fine class, i.e. into the product or into the coarse class (tailings), and will then be recycled into the mill. This probability is demonstrated by the Tromp curve dealt with in Chapter 9.

In Fig. 10/3 Tromp curves of a high quality separator with logarithmic abscissae are presented. Curve a represents the Tromp curve of the same classifier with a small circulating load; curve b with a great one; c is the ideal process. The usual bending upwards of curves a and b in the range of very fine particles indicates the deterioration of the separating effect. An explanation for this will be given in Chapter 18.

In Fig. 10/4 the calculated particle size distributions of the classified products in the RR chart are plotted. The milling product can be characterized by

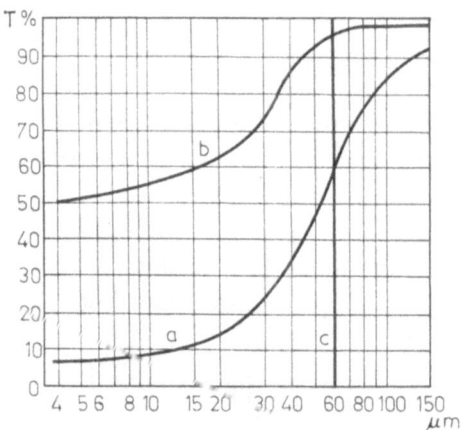

Fig. 10/3. Air separator Tromp curves
a small load; b great load; c ideal process

$n = 1$, $R(0.09) = 30\%$, it was ground by avoiding the range of agglomeration and separated according to curves a, b and c in Fig. 10/3.

In conformity with formula (9.1a) the ordinates of the Tromp curve can be calculated as

$$T = \frac{\Delta G}{\Delta D} \tag{10.7}$$

Table 4
Plotting of the Tromp curve

x μm	1	3	4	5	7.5	10	15	20
$g\%$		100		99.7		99		96.8
$d\%$		96.8		94.5		89		78
$p\%$		93		88.5		77		55
U		2.2		2.2		2.2		2.2
$\Delta p\%$	7		4.5		11.5		22	
$\Delta g\%$	—		0.3		0.7		2.2	
$1.2\Delta g$			0.4		0.8		2.6	
$\Delta d\%$	3.2		2.3		5.5		11	
$2.2\Delta d$	7		5		12		24	
$T\%$	---		8		7		11	

x μm	30	40	50	60	80	100	150	200
$g\%$		90		79		58		18
$d\%$		63		49		34		10
$p\%$		30		13		5		0.1
U		2.2		2.2		2.2		2.2
$\Delta p\%$	25		17		8		5	
$\Delta g\%$	6.8		11		21		40	
$1.2\Delta g$	8		13		25		48	
$\Delta d\%$	15		14		15		24	
$2.2\Delta d$	33		31		33		53	
$T\%$	24		43		76		91	

and with regard to

$$UP = D, \quad (U - 1)P = G, \quad UP\Delta d = (U - 1)P\Delta g + P\Delta p$$

$$T = \frac{U - 1}{U}\frac{\Delta g}{\Delta d}. \tag{10.8}$$

A numerical example for the calculation based upon formulae (10,4) and (10,8) is given in Table 4, giving values in conformity with curve a in Fig. 10/3 and the size distribution in Fig. 10/5. In the latter the full lines represent values from the sieve test, the dotted lines the extrapolation according to the constancy of U for the whole range.

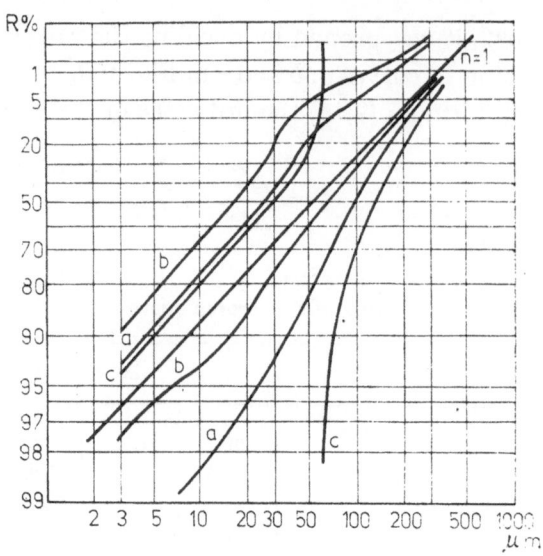

Fig. 10/4. Separation products of a separator feed (mill throughput) characterized by
$R(0.09) = 30\%$, $n = 1$

a, b, and c referring to Tromp curves in Fig. 10/3

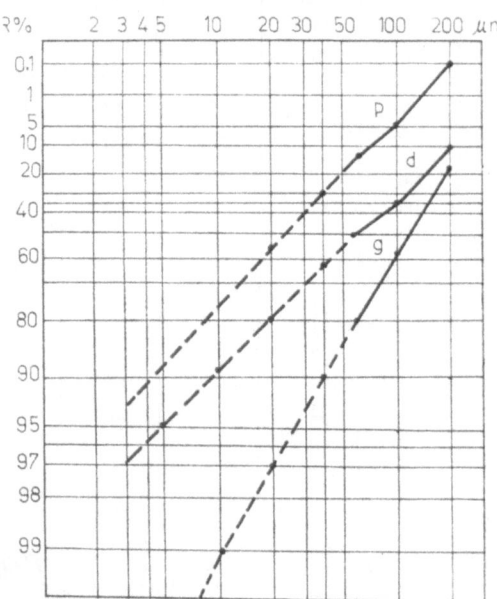

Fig. 10/5. Particle size distribution of separation fractions in closed circuit cement grinding
d throughput (separator feed), p fine product, g tailings. Separation process according to line a in Fig. 10/3

This numerical example is identical to the separation represented in Figs
9/2–9/4, so a comparison of both methods is possible. The value of the equalizing parameter of the Tromp curve can be read on Fig. 9/4 or Fig. 10/3 at
$100 - 100/U = 55\%$ as about 63 microns.

These somewhat lengthy calculations can be performed by the application of matrix algebra — which can be computer-programmed (Broadbent and Callcott 1956, Brown 1959).

Classification is described by a **Q** diagonal matrix. The elements in the diagonal are the probability values of getting into the fine class, i.e. 1-T, all other ele-

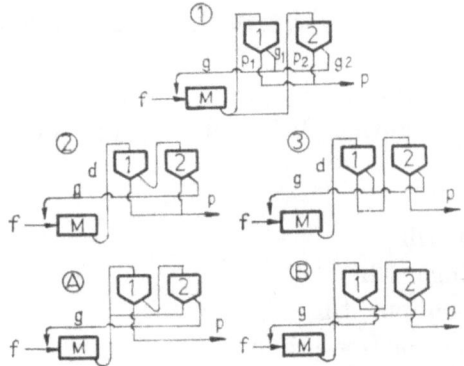

Fig. 10/6. Two separators in various connections

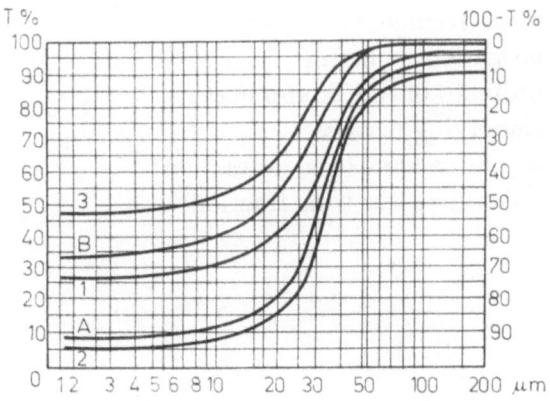

Fig. 10/7. Tromp curves referring to Fig. 10/6

ments are equal to zero. The mill product to be separated is represented by the column vector **d**. The classified fine product is given by the column vector **p**, where

$$\mathbf{p} = \mathbf{Qd}. \tag{10.9}$$

High capacity grinding mills are usually operated in a closed circuit with two identical classifiers in parallel installation (Fig. 10/6, Pos. 1). Commercial quality classifiers have Tromp curves indicating the recycling of a considerable part of fine particles, resulting in a decrease in the mill capacity. This can be avoided by connecting the two classifiers in series, directing the tailings of the first classifier into the second (Pos. 2). On the other hand if very strict fineness prescriptions are to be satisfied, the fine particles of the first classifier can be directed into the second one (Pos. 3). There is also the possibility to recycle either the fines or the tailings of the second classifier into the first one (Positions A and B respectively). These installations can be calculated by repeated application of formula (10.9) resulting in the modification of the Tromp curve (Fig. 10/7).

The problem of the cement grinding process with several classifiers has been dealt with in detail by the author (Beke, 1972) and by Jäger (1976).

Symbols in Chapter 10

F	feed, t/h
G	tailings, t/h
D	throughput, t/h
P	output or fine product, t/h
U	circuit coefficient
p, d and g	the respective sieve residues
h	separation size, microns
R	sieve residue
c	coefficient, t/h, μm
T	ordinate of the Tromp curve
$\mathbf{Q}$	classifying matrix
$\mathbf{d}$	column vector (coarse fraction)
$\mathbf{p}$	column vector (fine fraction)

CHAPTER 11

PRACTICABLE METHODS OF FINE GRINDING

Up to date technology often requires grinding fineness getting into the range of aggregation or even agglomeration (for definitions, see Chapter 4). As has been dealt with in the preceding chapters, to avoid or at least to retard agglomeration (with special regard to ball mills), there exist the following possibilities:

(a) small mill diameter

(b) grinding bodies not surpassing the minimal size to give rise to breakage

(c) short grinding time

(d) application of grinding aids.

Item (a) excludes large grinding capacities; item (b), involves the reduction of grinding body size, is not possible in conventional mills. The lowest grinding body size, when standard diaphragm slots are used, is limited to 15–20 mm, which corresponds according to formulae by Bombled or Bond (discussed in Chapter 13) to $x_{80} = 0.36$–0.64 mm or $R(0.09)$ greater than 50%. For particles of the size 0.03 mm grinding bodies of about 5 mm are convenient. This indicates that conventional ball mills—in avoiding agglomeration—are unable to produce other than medium fineness. Some years ago the firm F. L. Smidth (Copenhagen) introduced the "minipebs" mill where discharge is achieved without slotted diaphragms, small size grinding bodies of 4–8 mm can be used, in consequence of which very low values of sieve residue can be obtained (Cleemann 1972).

A short grinding time (item (c)), i.e. low number of impacts, causes a coarse product in any case. If, however, a closed-circuit grinding system is used and the coarse particles are repeatedly ground, we arrive at the nowadays mostly applied solution where the zone of agglomeration is largely avoided.

Item (d), the application of grinding aids, was dealt with in Chapter 7. Dependent upon the properties of the feed, aggregation can be imminent in an early stage of the grinding process, in which case methods (a), (b) and (c) must be combined with grinding aid dosage, as is done, for example, with cement grinding in the cited minipebs mill.

Problems of very fine grinding will be discussed in Chapter 16.

CHAPTER 12

TUMBLING MILL MECHANICS

A tumbling mill is a collective name for the generally known ball mills, rod mills, tube mills, pebble mills and autogeneous mills. For all these kinds of mills the mechanics can be dealt with together, there being no substantial difference in the grinding process.

There are two kinds of grinding body movements: either they describe an approximately parabolic trajectory, and knock against the material bed in a process known as cataracting; otherwise they slide and roll downwards on the material bed surface, this is known as cascading. The circumstances of these combined movements are very complicated and in addition to being influenced by gravitational and centrifugal forces and by friction, they are also influenced by the mutual effect of the lining of the mill and by the grinding bodies. The problem has been under investigation since the beginning of our century without, however, all aspects being completely elucidated.

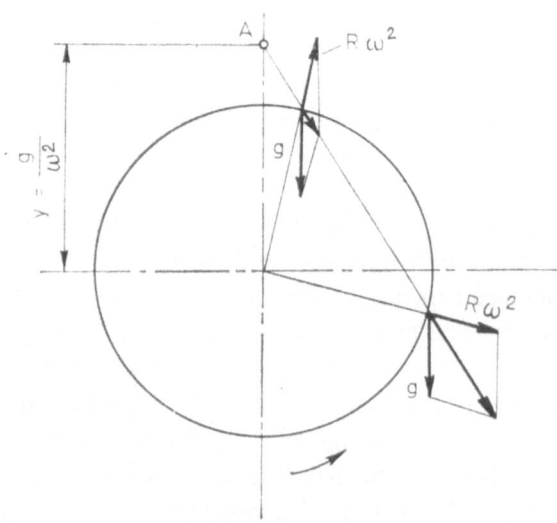

Fig. 12/1. Resultants of gravitational and centrifugal forces

The movement of the grinding media which rise together with the rotating mill shell and which later describe a parabolic trajectory without disturbing each other's motion was established by Davis (1920). This is a simplified method which offers an acceptable approach and is the one that is mostly applied. The above notwithstanding, there is, however, a mutually disturbing effect of the individual balls; this process was investigated by Joisel and Birebent (1951). The sliding-rolling movement of the grinding bodies as well as the sliding backwards of the whole charge as one block was written up by Uggla (1930).

The resultants of the gravitational and centrifugal forces form a centrifugal force field with the centre being at a distance of $Y = g/\omega^2$ above the mill axis (Fig. 12/1).

If $Y = g/\omega^2 = R$, the gravitational and centrifugal forces are in equilibrium at the top of the mill shell. If they reach this point, the grinding bodies adhere to the mill shell. This is the so called critical angular velocity or critical revolutions per minute (rpm or Hz)

$$\omega_{kr} = \sqrt{\frac{g}{R}}; \quad n_{kr} = \frac{42.3}{\sqrt{D}} \text{ rpm or } \frac{0.7}{\sqrt{D}} \text{ Hz} \tag{12.1}$$

where D is the inside diameter in metres, and R is the radius of the mill shell. Where $n > n_{kr}$, basically no ball movement can occur which therefore means that grinding cannot take place; this holds true unless there is an unwanted sliding of the balls.

The operational rpm is usually given as a percentage of the critical rpm.

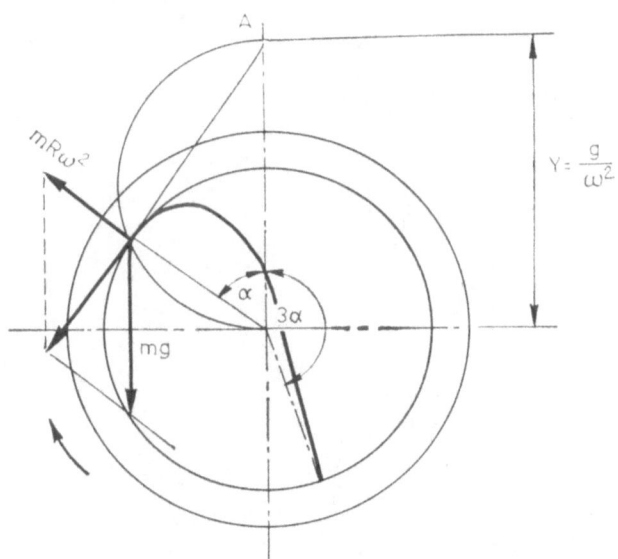

Fig. 12/2. Cataract grinding, parabolic trajectory

When the angular velocity is lower than the critical value, in the lower part of the shell the resultant force will show outwards, in the upper part inwards, its direction will change at the tangential point (Fig. 12/2). At this point the resultant force is tangential too, having no radial component. Evidently this is the starting point of the trajectory. Starting here tangentially the moving body will, with an acceptable approach, describe a parabola and will be projected to the opposite lower part and be dashed against the material and grinding body bed. Characteristic of the trajectory are its central angles: starting at $-\alpha$, and dashing at 3α the trajectory spans an angle of 4α.

For bodies (mostly balls but pieces of material too) moving on different radii the starting point of the trajectory is to be found on a semi-circle above the distance $Y = g/\omega^2$, the dashed point on a spiral line (Fig. 12/3).

For bodies elevating with the mill shell the maximum drop height can be attained at the speed

$$n = \frac{32}{\sqrt{D}} \text{ rpm} \left(\frac{0.53}{\sqrt{D}} \text{ Hz}\right) \tag{12.2}$$

which is 76% of the critical rpm. In this case the drop height is

$$H = Y = \frac{D}{2}\sqrt{3} \tag{12.3}$$

and $\alpha = 55°$. As is well known this is the usual operational speed of ball mills.

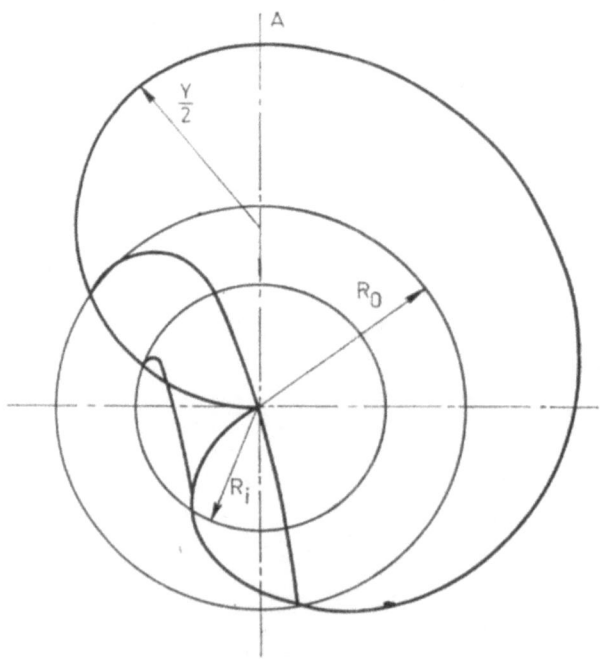

Fig. 12/3. Outer and inner parabolic trajectories

In the most inner trajectory the spiral and this trajectory have a common tangent. A trajectory on a smaller radius is impossible, the space being engaged by outer balls with greater energy (Fig. 12/3).

For a cataracting function to come into being the elevating grinding bodies must reach above mentioned starting points of the trajectories. In the sub-critical range this is impossible for one single ball since on reaching the angle of friction it rolls down.

The free surface of the mill charge — consisting of a multitude of grinding media — forms an equipotential surface perpendicular to the force field. Friction neglected it is a circular cylinder with its axis in the centre of the force field. The effect of friction deforms this cylindrical surface, its angle normal to the force field becomes equal to the angle of friction ϱ (Fig. 12/4). Such a curve, the generatrix of the cylinder, is a logarithmic spiral. Its equation in the polar coordinate system is (Uggla 1930)

$$r = ae^{\mu\varphi} \tag{12.4}$$

where r and φ are polar coordinates, a is a constant, and μ the friction coefficient).

The mill charge has, according to Fig. 12/4, two parts separated by the equipotential surface: the lower part elevates in relative resting state together with the mill shell, the upper part rolls (cascading) downwards, the upper boundary of this part is a plane with the slope ϱ. This lower part of the charge

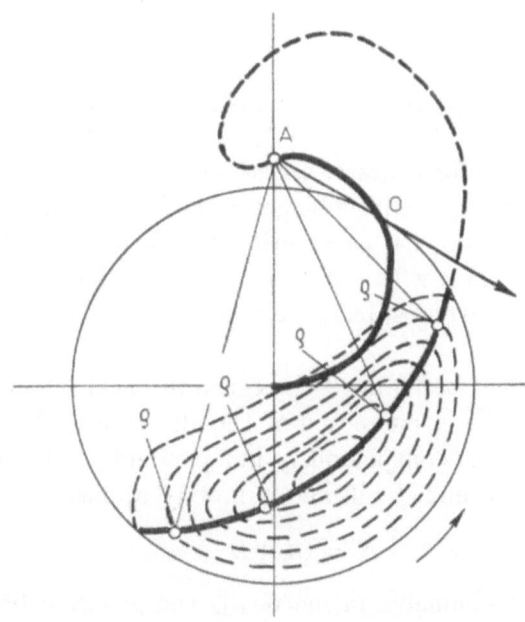

Fig. 12/4. Cascade grinding

displays a fine grinding effect too: affected by the mill shell the individual balls will rotate (Fig. 12/5); this effect is brought about by the friction to the mill shell.

To summarize: a cascading motion is always taking place. The condition for partial cataracting is a sufficient grinding body charge and a suitable friction coefficient. This latter can be influenced by the shape of the liner plates.

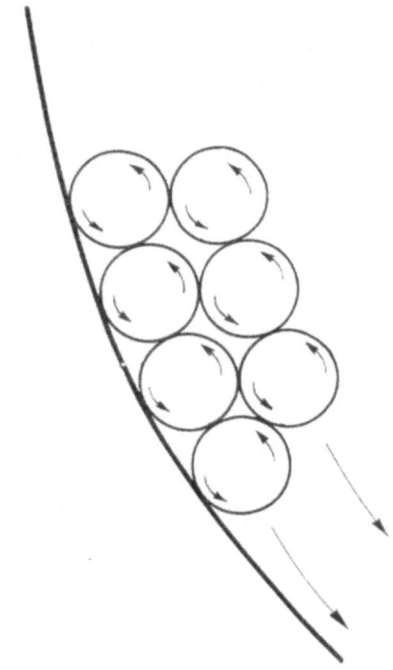

Fig. 12/5. Spin movement of grinding bodies

Cataracting is desired for coarser, cascading for finer grinding; in conse·quence of this for coarser grinding a bigger mill charge is necessary. As for the shape of grinding bodies: with cataracting the weight effect is well accomplished by spherical bodies; with cascading a large cylindrical surface is effective. So for cataracting purposes, balls are more suitable, whereas for cascading, cylpebses are better.

The energy demand of ball mills — disregarding the mechanical losses (bearing-friction, mill drive) — is composed of the elevating and accelerating work of the mill charge; this is conspicuously independent of the mill throughput.

The semiempirical formula by Blanc is mostly applied:

$$N = CQ\sqrt{D} \tag{12.5}$$

D being the mill shell diameter in metres, Q the grinding body charge in t, C a constant which includes the correction due to the material particles in the free space between the grinding bodies. The value of C is strongly dependent on the filling rate ε. In the case of a large filling rate, the elevating height of the charge is smaller. Values of C are presented in Table 5. With these values we get N in HP. In Fig. 12/6 values of $0.736\,C$ are presented. Using these values for calculating, we get N in kW.

Table 5

Values of factor C in HP/t $\sqrt{m}$

Filling rate %	10	20	30	40	50
Large steel balls	11.9	11.1	9.9	8.5	7.0
Little steel balls or cylpebs	11.5	10.6	9.5	8.2	6.8
Flintstone or porcelain balls	13.3	12.2	11.1	9.5	7.8

Formula (12.5) can be written

$$N = C\,\frac{D^{2.5}\,\pi}{4}\,L\varepsilon\delta_c \qquad (12.5\text{a})$$

where ε is the filling rate, δ_c the volume density of the charge.

It is important to realize that the effectivity, i.e. the specific energy consumption of a ball mill, improves with increasing shell diameter so the output in t/h is proportional to $D^{2.6}$–$D^{2.7}$ instead of $D^{2.5}$, and with autogeneous mills even $D^{2.8}$–D^{3}.

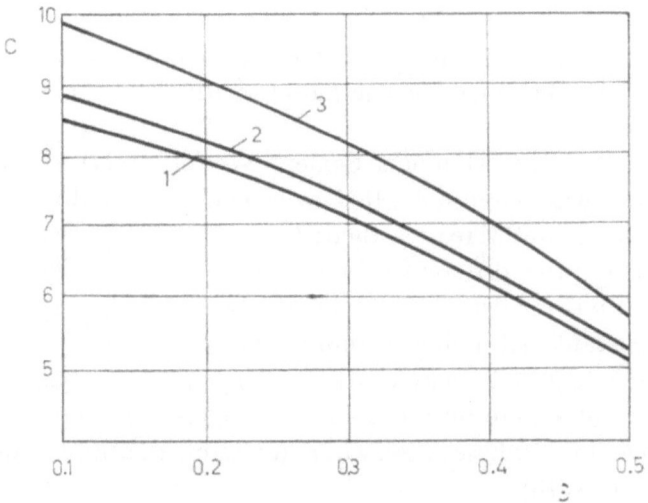

Fig. 12/6. Values of C in formula (12.5), kWh/t $\sqrt{m}$
1 small bodies; 2 big balls; 3 flint stone or porcelain balls

To estimate savings in specific energy consumption, Rowland (1972, 1975) proposed a correction factor $(2.44/D)^{0.2}$ which is valid as far as $D = 3.81$ m $= 12.5'$, where the value 0.914 acts as a stabilizer.

It is of interest that the rpm is not included in formula (12.5). The operational rpm is generally scarcely different from (12.2); with smooth liners in the numerator we can find 33, with wave-shaped or corrugated shell liners 30. For wet grinding, because of slippage of the charge, 10% can be added. Commercial high capacity ball mills are driven mostly by synchronous or asynchronous electric motors with intermediate girth ring and pinion or central drive; this

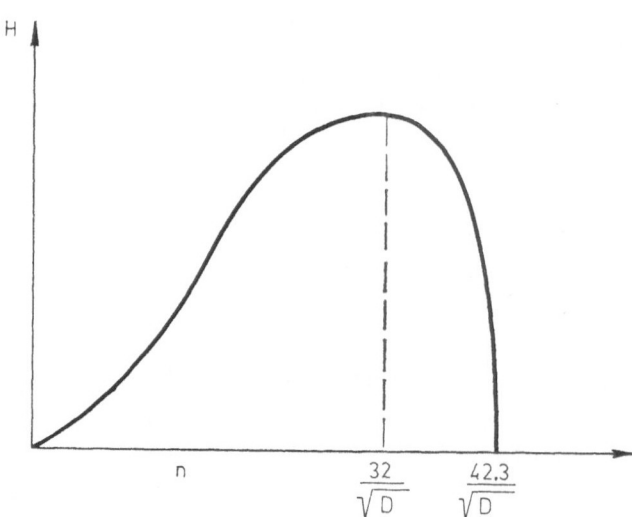

Fig. 12/7. Drop height against rpm

means that no alteration in the rpm is possible. The newly introduced gearless mill drive by low frequency electric motors with thyristor control allows an rpm regulation.

In this respect in the Guidonia Cement Works (Italy) a number of tests were performed with large 4.8×15.4 m cement mills (Olivero et al. 1977). Various behavior modes were investigated for a range of critical speeds between 66 and 79.8% and two different grinding media charges of 32 and 28%. The tests showed that smaller filling rate requires a higher speed of rotation — a very important factor since it represents the possibility to neutralize the wear of the grinding body charge witout stoppage of operation.

Analogous results were obtained in the Belgorod cement factory (USSR) using a 2×10.5 m mill whose speed was regulated with a special device (Kovaljukh and Gud, 1978).

As an explanation, Fig. 12/7 presents the grinding body fall height H as a function of rpm. A flat maximum is indicated at $32/\sqrt{D}$.

Combined with formula (12.5) it can be stated that in a very small range of speed variation there exists a proportionality between energy demand and rpm influenced by the grinding body charge too.

For the energy demand, instead of (12.5), Bond (1961) gave different empirical formulae for wet rod mills and wet and dry ball mills, where the diameter occurs as $D^{0.4}$. The formula contains — in accordance with above explanation — the factor Cs (fraction of critical speed) and the filling rate of the grinding media charge.

It is important to stress, however, that a simultaneous slippage-free mill charge movement was supposed. With smooth liners and especially with wet grinding this is not the case. So, for example, in his cited paper Uggla (1930) described and explained periodical backwards slippage of the charge. To avoid this phenomenon corrugated or shiplap liners are utilized. This problem was dealt with in particular by a paper by Halbart and Freymann (1955–1956).

Taking also the liner wear into consideration the problem was thoroughly investigated by Eifel and Schönert (1976).

Analysing the problem in the opposite direction Hukki (1958) developed quite another process: allowing deliberately the mill charge to slide back in one block, grinding takes place by friction. The shell rpm surpasses considerably the critical; inside the shell radial comb shaped plates are welded. Between these plates a "cushion" of material is formed so by autogeneous grinding no metal wear is to be feared. The method is now being developed.

The passing of material along the mill takes place because of the level difference between the inlet and shell bottom. Ventilation of the mill space enhances this material movement. Air swept mills naturally furnish coarser products: particles equal or less than the size conforming to the air velocity are taken away without further grinding. In multichamber mills of large L/D ratio, double wall partitions with elevating plates are used, in this way the level difference is restored, the throughput capacity can be increased.

The holdup and retention time in the mill are interesting characteristics. In the case of the analogous void factor of the grinding body charge the holdup is proportional to the mill volume or D^2L. The retention time is the quotient of holdup and output, this latter being taken as proportional to $D^{2.5}L$. Thus, we get the relationship that retention time is inversely proportional to $\sqrt{D}$; it is conspicuously independent of mill length.

Calculating with a specific energy consumption of 35 kWh/t and normal void volume of the charge we get $50/\sqrt{D}$. But this is evidently valid only for open circuits. In a closed circuit the energy is consumed in multiple passages; for a single passage we get the retention time by dividing by the circuit coefficient U according to formula (10.2).

One must not forget that the mass of holdup is influenced by the void volume in the grinding body charge, i.e. its porosity. Wear decreases the porosity thereby promoting the choking of the mill. With bigger diameter the retention time decreases, the product gets coarser; in closed circuit the recycling increases, in extreme circumstances choking of the mill can occur. Thus, in high capacity mills a charge of high porosity, i.e. a uniform grinding body charge, is required. But the increase of porosity is limited by the laws of geometry so capacity increase can be limited by these circumstances too. Present day mills are, however, below this limit.

Heat generation is an important phenomenon in the mill, especially with dry grinding. The heat is carried away by the milling product itself, by the airstream and by the heat transfer of the mill shell.

According to formula (12.5) the heat development is proportional to $D^{2.5}$ with L/D being constant even with $D^{3.5}$; whereas the cooling surface is proportional to D^2. Therefore with increasing grinding capacity the temperature in the mill space or that of the product will increase too. This phenomenon becomes important in the case of cement grinding were the product temperature is limited to about 110 °C, above this temperature gypsum loses its hydrate-water causing the false setting of cement.

To control the function of ball mills the so called "mill diagram" is utilized as it illustrates the decrease of sieve residue along the mill. In a series of recent investigations carried out by Opoczky (1978) the change of the uniform-

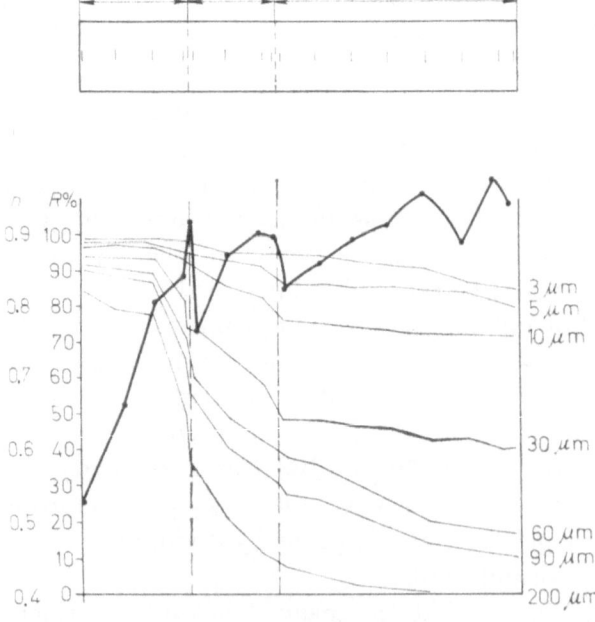

Fig. 12/8. Mill diagram

ity coefficient, according to formula (3.9a), was determined too, this latter indicates the phenomenon of aggregation. It is interesting to note that near to the partition walls a jam produces aggregation (Fig. 12/8).

Symbols in Chapter 12

D	inside diameter of mill shell, m
R	radius of mill shell, m
L	mill length, m
n	mill speed, rpm or Hz
ω	angular velocity 1/s
Y	distance of force field centre, m
H	drop height, m
μ	friction coefficient
ϱ	angle of friction
a, C	constants
N	energy demand, kW or HP
Q	mill charge, t
ε	filling rate (percentage)
δ_c	charge volume density, t/m³
Cs	percentage of critical speed

CHAPTER 13

GRINDING BODY CHARGE AND WEAR

The economical functioning of ball mills necessitates the application of the most suitable grinding body charge, a task of the plant management and not of the machine supplier. Here, the following questions need to be clarified:

- — mass of mill charge
- — size of the largest and smallest grinding body
- — size distribution of the grinding media
- — grinding media wear.

The mass of mill charge, the so called filling rate, is given as a percentage of the inner mill volume. As seen in the preceding, for cataracting a bigger, for cascading a smaller filling rate is suitable. Cataracting is the method for coarse grinding, cascading for finer grinding. In multichamber mills the coarse stage chambers obviously require the larger filling rate.

Filling rates in three-chamber open circuit mills (according to Anselm) are as follows:

First chamber	30%
Second chamber	27%
Third chamber	20–24%

In closed circuit grinding a coarser mill product is required and as a consequence, cataracting is overwhelming: the filling rate will be 31–33%; in air swept mills it will be 26–32%. A larger filling rate naturally increases the capacity per mill volume.

In formula (12.5) for the energy demand, the mass (weight) of the charge is included. To determine this mass, one needs to consider the following factors: The balls are equally sized and it is supposed that the free volume between the balls is independent of their size; in cubical arrangement the balls occupy $\pi/6 = 52\%$ of the space, with a porosity of 48%; by intensive shaking a more dense arrangement can develop; in a prismatic net the rate of occupied space is 60%; in a pyramidal net 74%.

With cylindrical grinding media (cylpebses or rods) the rate of occupied space is larger: in a regular net it is $\pi/4 = 79\%$, in a triangular net 83%. However, in operational conditions because of the continuous movement of the grinding bodies, the most dense arrangement cannot develop.

Although the percentage of porosity is independent of ball size, the sizes of the pores are not without effect on the grinding process.

High grinding fineness requires smaller pores to hinder the throughflow of coarse particles. This is the case with open circuit grinding where the grinding fineness comes into being in the mill.

On the other hand in closed circuit operation the product fineness is determined by the function of the classifier, here a large throughflow (see formula (10.4) of the circuit coefficient) in the mill is required. This can be ensured by larger pores, i.e. bigger ball sizes. In closed circuit grinding cylpebses are not allowed.

The charge does not consist of equal-sized balls and even balls which were originally equal in size become smaller as a result of wear.

Practical data for the specific mass (weight) per volume or volume density are

larger balls (60–100 mm)	4400–4600 kg/m³
smaller balls (30–40 mm)	4600–4900 kg/m³
cylpebses	4600–4800 kg/m³
flint stone, porcelain balls	1650 kg/m³.

In quick calculations we can take for the whole mill an average of about 4600 kg/m³.

The purpose of *ball size selection* is to provide sufficient kinetic energy to bring about breakage.

For simplicity we suppose spherical grinding bodies (i.e. balls) and a kinematic similarity, i.e. the same percentage of critical rpm. In this case the kinetic energy will be proportional to Dd^3. A formula of general validity will then have the form

$$Dd^3 = f(x) \tag{13.1}$$

where D is the mill shell, d the ball, and x represents the feed size.

The most simple rule will yield the classical volume theory by Kick (Chapter 1), i.e. an energy demand proportional to the particle volume

$$Dd^3 \sim x^3 \tag{13.2}$$

Papadakis (1960) proposed

$$Dd^3 \sim x^2 \tag{13.3}$$

calculating with the force and not the energy of breakage.

A more accurate formula by Bombled (1967) is

$$d = 93 \frac{x_{80}^{0.4}}{\sqrt[3]{BD}}$$ (13.4)

x_{80} and d being in mm, D in m, and B the grindability index (for cement clinker $B = 50$–100 cm²/J). In other sources we find lower B values resulting in larger ball sizes.

The most frequently used formula was proposed by Bond (1958)

$$d = \sqrt{\frac{x_{80}}{K}} \sqrt[3]{\frac{\delta Wi}{Cs \sqrt{D}}}$$ (13.5)

where d and D are in inches, x_{80} in microns, Wi is the work index, δ the specific weight or density (for cement clinker 13.5 or 3.1 respectively), K for dry grinding with steel balls 335, Cs the percentage of the critical rpm. In the metric system we get d in mm by multiplying the result by 20, D is then substituted in m. It is conspicuous that in (13.5) we see $d^3 \sqrt{D}$ instead of $d^3 D$, since Bond attaches less importance to the mill size than does Bombled.

A very frequent practical formula, disregarding the mill size, is (Starke 1935)

$$cx_{80} = d^2$$ (13.6)

and for cement grinding with $c = 625$

$$d = 25 \sqrt{x_{80}}$$ (13.6a)

with d and x_{80} being in mm.

Formula (13.6a) provides somewhat bigger ball sizes than the preceding. So, for example, for $D = 2$ m and $x_{80} = 10$ mm we get, according to (13.4) (13.5), and (13.6a): $d = 50$, 55, and 80 mm, taking $B = 50$ cm²/joule in (13.4). At the mill outlet for $x_{80} = 0.03$ mm the ball size is 5,3 or 4 mm.

The calculated values represent minimum ball sizes capable of executing breakage. It can to be seen that balls of the size calculated for the mill outlet are not applicable in conventional mill constructions because of the diaphragm slots of 12–14 mm.

What occurs then if the ball sizes exceed the calculated sizes? As was shown in Chapter 4, energy overdose brings about agglomeration. And taking it into consideration that in all sections along the mill, even in the coarse ranges, there come into being—in accordance with the laws of particle size distribution—very fine particles too, we can establish the following most important fact: *in ball mills, partial agglomeration is unavoidable* even in the beginning phase of the grinding process.

As for the *size distribution of the grinding media*, this must be in accordance with the decreasing material size. This latter is usually presented in the so-called mill diagram showing the decreasing sieve residue along the mill. For our purposes a diagram showing the decreasing material size characterized by either $\bar{x}$ or x_{80} will be more instructive.

According to Rittinger's law the decrease of material size can be described as $\bar{x} = c/l$, where l is the distance from the mill inlet. This is of course a simplified approach in that the energy supply to the individual mill sections does not remain constant along the mill due to different filling rate, liner shape and friction coefficient.

Applying either formula (13.5) or (13.6), the ball size along the mill can be described as $d = c/\sqrt{l}$, a hyperbola of higher order as seen in semilogarithmic graphical representation in Fig. 13/1.

This figure, however, supplies data far from customary ball size selections. this is because the supposed Rittinger law does not conform with the real process, and what is more important: the ball size is limited downwards by the size of the diaphragm slots, i. e. at about 15–20 mm.

Bombled, in his cited paper, supposes a linear increase of the Blaine surface along the mill and with empirical correction of the Rittinger equation, he arrives at the formula

$$x_{80} = \frac{x_f}{kl^2 + 1} \tag{13.7}$$

where x_f is 80% passing size of the feed, k is a constant characterizing the equip-

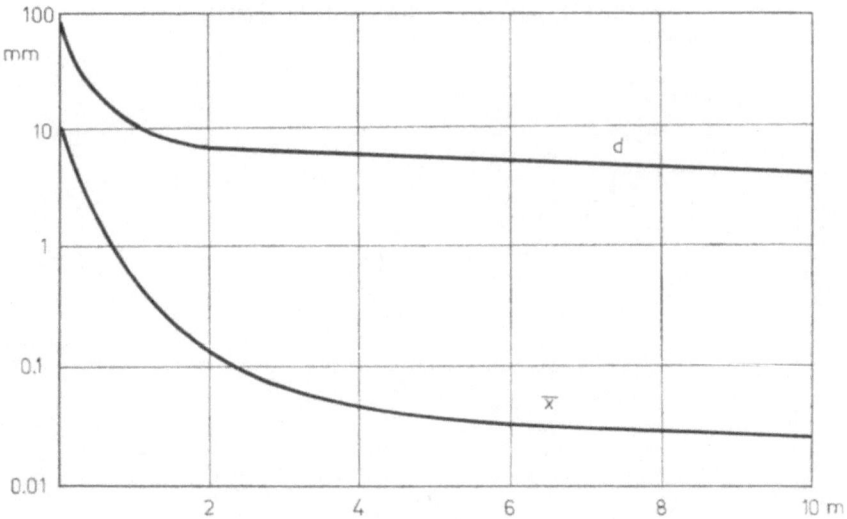

Fig. 13/1. Corresponding size reduction values of grinding body and feed along the mill

ment and ground material. The value of d can be then calculated by formula (13.4). A further factor is that the resulting curves seem too steep.

Based on model experiments with 1.2×2.5 m mill Schramm and Gaitsch (1974) observed an exponential law of the material as well as of the ball size decrease, in semilogarithmic scale they obtained straight lines as shown in Fig. 13/2. Their formula is

$$d = d_k e^{-gl} \tag{13.8}$$

where d_k is the starting "imaginary" ball size calculated according to formula (13.5), and g is constant. The real ball size distribution is of course stepwise. If it is supposed that there is a uniform ball charge along the mill the mass of $70, 60 \ldots 18$ mm balls is proportional to the respective mill lengths.

The grinding bodies of different sizes must be distributed along the mill in conformity with the decreasing material particles. To serve this purpose two methods are known: multichamber (or compound) mills — those divided with one or more partitions (diaphragms) into two or more compartments, and/ or conical shaped classifying lining plates. The partitions are designed to allow ventilating air and undersize particles to pass into the next chamber and to prevent access of oversize balls. Thus the area of the diaphragm slots has a significant influence upon the function of the mill, the free area must be at least 10% of the total mill cross section.

Usual compartment lengths with three-chamber compound mills, where the ratio $L/D = 5$—6

1st chamber   20% mill length
2nd chamber   30% mill length
3rd chamber   50% mill length

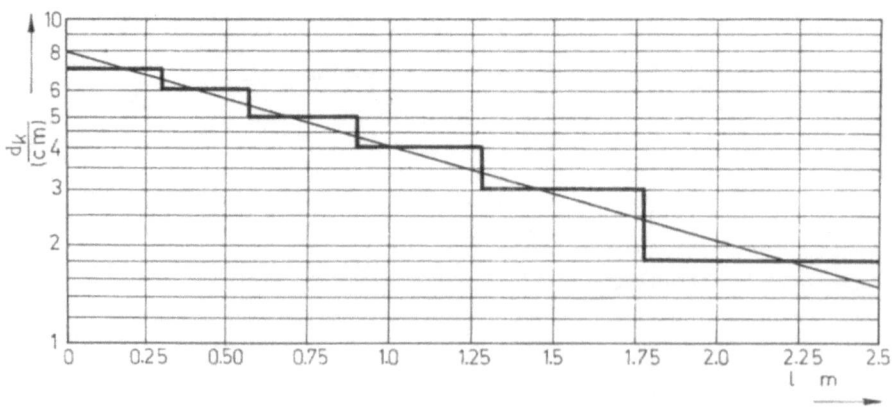

Fig. 13/2. Change of grinding body size along the mill. (Investigations by Schramm and Gaitsch)

In a closed circuit system we allow a coarser mill product with the usual ratio $L/D = 3$—4, two chambers are sufficient, the fine grinding chamber occupies about two third of the mill length.

As for classifying liners, Slegten (1964), the top expert on this problem recommends their application only into the fine grinding chamber. The explanation of this recommendation is that classifying plates drive the big balls toward the feed end exerting a harmful pressure on the front walls. According to experience coarse grinding requires grinding media of different sizes and furthermore large pieces of the feed are driven towards the partition wall, they accumulate there without rupture.

Into the pregrinding chamber, according to Slegten's view, a mixture of 3 or 4 ball sizes (e.g. 90, 80 and 70 mm) in equal number is suitable. On the other hand, the number of balls in the fine grinding chamber is increased but the size of these balls is decreased, in accordance with the increasing surface of the ground material.

With regard to the size distribution of grinding media, as mentioned above, very differing views are expounded in the literature in view of which it seems impossible to give an unambiguous directive. As an interesting example, we present here the system elaborated in three different versions by Kemmann (1972). He composed a table of ball sizes decreasing in 10 mm steps from 100 mm, and cylpebs sizes decreasing in 2 mm steps from 38 mm downwards. For all sizes he calculated the G weight and S surface and for the effectivity three index numbers:

Index A: according to G/S
Index B: according to S/G
Index C: according to the difference between two neighbouring sizes.

Taking $\Sigma G/S = 100\%$ resp. $\Sigma S/G = 100\%$ we get:

	weight	number	surface
According to A	$\dfrac{G/S}{\Sigma G/S} = \dfrac{G}{S}$	$\dfrac{1}{G}\dfrac{G/S}{\Sigma G/S} = \dfrac{1}{S}$	$\dfrac{1}{\Sigma G/S} = 1$
According to B	$\dfrac{S/G}{\Sigma S/G} = \dfrac{S}{G}$	$\dfrac{1}{G}\dfrac{S/G}{\Sigma S/G} = \dfrac{S}{G^2}$	$\dfrac{S^2}{G^2}\dfrac{1}{\Sigma S/G} = \dfrac{S^2}{G^2}$

The series according to C furnishes values between A_4 and B.

The series according to index A is characterized by the uniform surface of all ball size-fractions; ratio G/S decreases, in the beginning stage individual force is effective upon a larger surface, consequently with larger balls it is suitable for coarse grinding.

Series B is characterized by increasing G/S ratio, it is in conformity with increasing surface, suitable to fine grinding.

In series C a mixture of both preceding methods takes place, consequently it must be applied in closed circuit grinding which requires a medium grinding fineness.

Börner (1965) presented a comprehensive figure summarizing empirical data collected in the German cement grinding practice (Fig. 13/3). Sphere I–II is related to smooth, III–IV to classifying liners. In the figure the effect of grinding body wear (see equation (13.11)) and for comparison the for concrete preparation competent so called Fuller curve is included too, this latter gives the most dense arrangement — naturally to be avoided in mill charge. All these considerations provide guidance for the starting mill charge but its effectivity must be confirmed and, if necessary, modified by evaluation of plant experience (mill diagram, capacity and energy consumption measurements).

As for ore and coal grinding with moderate grinding fineness (some tenths of millimetres) the L/D ratio is about one, the starting charge frequently consists

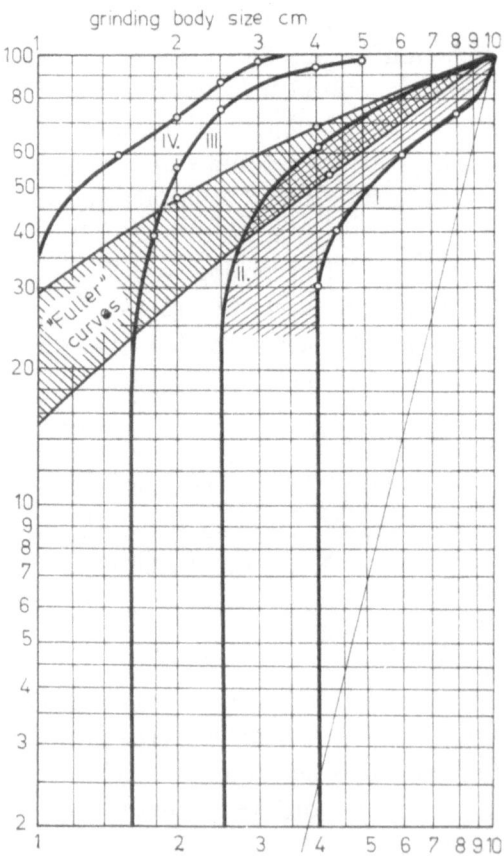

Fig. 13/3. Practical grinding body sizes for cement grinding according to Börner. I–II cylindrical, III–IV classifying liners

of uniformly sized balls in accordance with the size and grindability of the feed. Davis (1920) also expounded this view in his classical paper.

Having succeeded, however, in finding the effective starting charge the wear will modify (make worse) the ball size distribution. Thus, it is very important, and economical, to use a wear-resistant mill charge but, following the inevitable wear — be it earlier or later, attempts must be made to restitute or at least approach the original ball size composition. This is our next task.

The problem of grinding media wear has been the subject of widespread literature. (This, too, was dealt with in the pioneer work of Davis.) As indicated above, Davis accepted the cataracting function and in his view the wear is proportional to the mass or weight of the grinding media. Davis considers the starting mill charge to be composed of uniformly sized balls (as is usual with short ore grinding mills) and later the missing mass is replaced with balls of the original size. So an equilibrium ball size distribution develops.

His demonstrations resulted in an equilibrium distribution in a parabola of the third order. If the worn mass is replaced continuously by d_m original sized balls, the fraction between d_a and d_b will be

$$\varDelta d = \frac{d_a^3 - d_b^3}{d_m^3} \qquad (13.9)$$

and if balls worn to d_0 are cleared away

$$\varDelta d = \frac{d_a^3 - d_b^3}{d_m^3 - d_0^3} \qquad (13.9a)$$

and the time of wear from d_a to d_b

$$t = \frac{3}{k} \ln \frac{d_a}{d_b} \qquad (13.10)$$

where k is constant characteristic to the grinding process and ball material.

The supposition of Davis, i.e. a neat cataract function and in consequence a mass proportional wear, cannot be accepted. The alternative view, a wear proportional to surface or d^2, supposes a neat cascade function.

Paulsen (1964) supposes a time-proportional diameter decrease

$$-\frac{dd}{dt} = c_1$$

and the change of D percentage passing size

$$-\frac{dD}{dt} = \frac{d^3 \pi}{6} \delta z \frac{100}{G}$$

where z is the number of replaced balls, δ the density and G the relative undersize mass in the mill charge. To explain this equation we must consider that

91

the decrease of passing mass by time is equal to the relative mass of the replaced balls of size d. The constant values collected

$$-\frac{dD}{dt} c_2 d^3$$

and t eliminated we get the equilibrium distribution as

$$\int_D^1 dD = c \int_d^{d_m} d^3\, dd$$

and

$$1 - D = c'(d_m^4 - d^4) \qquad (13.11)$$

giving, in a double logarithmic net, a straight line with the slope 4 (Fig. 13/3).

Neat cascading does not hold either, but a mixture of cataracting and cascading does. Therefore individual authors recommend exponents between 3 and 4. So we find, after Bond (1958), the exponent 3.84; after Perow and Brand (1954), 3.7.

In his cited paper Bond recommends as starting charge the equilibrium ball size distribution. This is in contradiction with the views of the last-cited authors. This proposition can be recommended for coarse grinding in short ball mills and is of course more favourable than a uniformly sized ball charge.

It is worth comparing Fig. 13/2 and Fig. 13/3. The one giving a straight line is semilogarithmic, the other (equilibrium state) is in double logarithmic chart; it is impossible to attain covering of both lines.

The *time function of wear* can be deduced either according to the cataract or to the cascade function (Bombled 1971, Bernutat 1964).

With cataracting, as seen above, the wear can be supposed as being proportional to the mass of the ball:

$$-\frac{dG}{dt} = cG$$

and

$$G = G_m e^{-ct} \qquad (13.12)$$

or

$$d = d_m e^{-c't} \qquad (13.13)$$

With neat cascading the wear is considered proportional to the ball surface

$$-\frac{dG}{dt} = cS = cd^2$$

and

$$d = d_m - ct \qquad (13.14)$$

The actual function is a mixture of cataracting and cascading. The ratio of the two functions is characterized by a and b. Bombled proposes the formula

$$d = \frac{a}{a+b}(d_m - k't) + \frac{b}{a+b}d_m e^{-k''t} \tag{13.15}$$

Bernutat, in his derivation, takes the mill shell diameter into consideration too so he arrives at the formulae

$$-dG = (k_1 G + k_2 S)\,dt$$

For cataracting

$$d = d_m e^{-k't\sqrt{D}} \tag{13.16a}$$

and for cascading

$$d = d_m - k''t\sqrt{D} \tag{13.16b}$$

According to (13.16a), with cataracting the influence of mill size to ball wear surpasses that of cascading. Considering only (13.16a) and expanding in series, calculating with mass instead of diameter we get

$$\Delta G = kG_m \sqrt{D}\,t$$

which is valid naturally for the whole charge too

$$\Delta Q = kQ\sqrt{D}\,t$$

By comparing the latter with formula (12.5) of the energy demand we arrive at the very important relation

$$\frac{\Delta Q}{Nt} = \text{const}$$

which means that specific wear in g/kWh is independent of mill size.

The finding of Bombled is quite the same. On analysing the influence of mill shell diameter he was able to state that if we calculate with ball diameters according to (13.4) the wear is proportional to $D^{0.16}$, practically independent of D.

The present trend of development requires ever increasing mill sizes and this trend is not disturbed by increasing grinding media wear.

Simultaneous cataracting and cascading was taken into consideration by Razumov (giving the exponent 2.3 — according to Perow and Brand (1954) loc. cit. p. 101) with exponent n. With cataracting $n = 3$, with cascading $n = 2$. The sequence of calculation is

$$-\frac{dG}{dt} = cd^n$$

or

$$-\frac{dd}{dt} = cd^{n-2}$$

and the time of size decrease

$$t = \frac{d_m^{3-n} - d^{3-n}}{c(3-n)} \qquad (13.18)$$

n is given by Razumov for ore grinding as 2.3; by Börner (1965) for cement grinding as 2.6–2.8; by Bombled as 2.5.

Kolostori (1975) made long-term observations concerning diameter decrease of different ball sizes in a commercial cement mill. Plotting a continuous line he got an empirical wear function (Fig. 13/4). Based upon these data he calculated the ball size composition of the first chamber charged initially with 90, 80 and 70 mm balls. Compensating the worn and missing part regularly with 90 mm balls he demonstrated the successive deformation and refinement

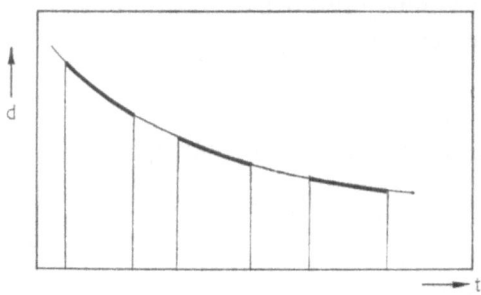

Fig. 13/4. Empirical wear function according to Kolostori

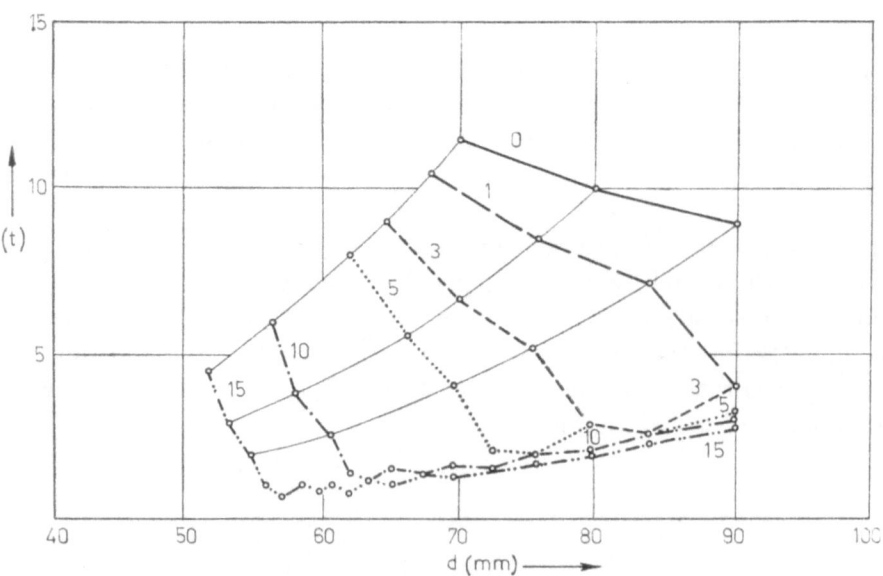

Fig. 13/5. Development of grinding body composition in case of weight compensation by biggest sized balls (Kolostori)

of the charge and its becoming unable to provide efficient grinding (Fig. 13/5). So, as generally known, the tumbling of the mill, classifying the worn charge and filling with a quite new charge is unavoidable. This is a very tedious and time-consuming task necessitating the stoppage of grinding for several days; it leads to deterioration in the availability factor, the very important index-number, the quotient of operating time to the sum of operating, and stoppage time.

To prolong the working period of the mill, wear-resistant grinding media must be utilized. In this respect, over the last years considerable results have been obtained. So, for example, in the cement industry of the FRG, the following progress was demonstrated (Drohsihn 1972):

before 1960	CMn steel	1000 g/t
	1966 special wear-resistant alloy	100 g/t
	1971 special wear-resistant alloy	50 g/t

Previously, manganese alloys were usual, today CrNi alloys with martensitic texture are required to attain the above favourable values.

The average data of grinding body wear in g/t as a function of Brinell hardness for portland cement, blast furnace portland cement and cement raw meal are presented in Fig. 13/6.

Detailed data on grinding body material composition, properties and related specific wear values can be found in Duda's Cement Data Book (1976–1977).

The total elimination of grinding body wear is possible by autogeneous grinding, where large pieces of the feed execute the grinding mostly by a cataracting movement. With no heavy grinding bodies in the mill, a greater cataracting height, i.e. bigger mill shell diameter, is required. In this way, we arrive

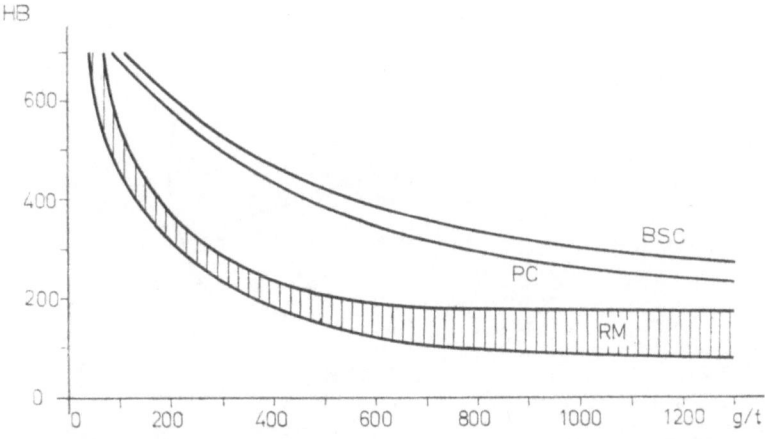

Fig. 13/6. Grinding body wear (g/t) against Brinell hardness in case of Portland cement, blast furnace slag cement and cement raw meal grinding

at the relatively short mills with an L/D ratio of 1/2–1/5. Nowadays, ore grinding autogeneous mills have the largest capacity ever attained reaching the driving power of 10 000–15 000 kW with diameters of 10–12 m. Of course the retention time is short, so only a moderate fineness can be obtained. (A mill of 12 m × 6 m results in about 20 000 kW.)

Autogeneous grinding has a selective effect, viz. heterogeneous materials will break along cleavage planes, components of different grindability can be separated by classification.

The application of autogeneous mills is advantageous if one of the following conditions holds:

— very hard, abrasive feed
— drying-grinding of feed of large moisture content
— selective grinding.

If the feed has an unsufficient content of large pieces, in the middle ranges a saturation takes place, and the mill will choke. In this case a 120–150 mm steel ball charge is needed with a filling rate of about 5% (semi-autogeneous grinding).

In his above cited paper Bernutat (1964) analyses the liner wear too. He proposes the following formula

$$\Delta M = kD^{1.5}t \qquad (13.19)$$

where D is the inside liner diameter (in present, worn state), which means that with progress of wear its velocity increases. With increasing D, both energy demand and output increase too.

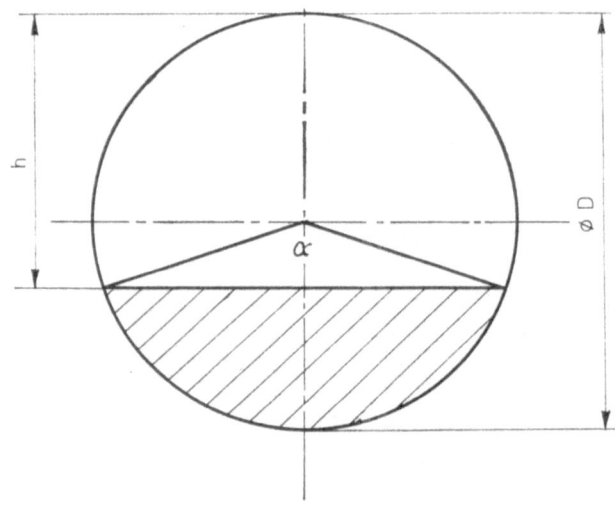

Fig. 13/7. Scheme for calculation of worn grinding body mass

The wear of grinding media requires replacing of the missing mass. A drop in electric power demand (indicated in kW or amp) — and of course a simultaneous decrease of output and efficiency — calls attention to the urgency to complete the charge. The mass of the shortage can be determined by the difference of the free height above the charge level in the starting and worn condition. For both cases ε filling rate is calculated according to Fig. 13/7 by the formulae

$$\varepsilon = \frac{\dfrac{\alpha}{2} - \sin\dfrac{\alpha}{2}\cos\dfrac{\alpha}{2}}{\pi} \tag{13.20a}$$

and

$$\cos\frac{\alpha}{2} = \frac{2h}{D} - 1 \tag{13.20b}$$

ε as a function of h/D is presented in Fig 13/8.

For a quick estimation, the counting of uncovered liner segments is sufficient.

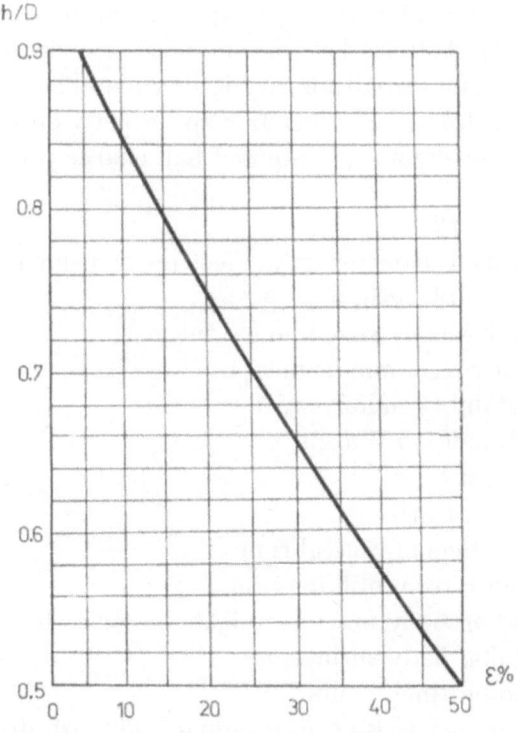

Fig. 13/8. Diagram for estimation of worn mass

In the preceding, the problem of mill charge was dealt with at considerable length, and by expounding the sometimes contradictory views of several authors.

To conclude this chapter, we cite here some sentences of an important paper by Börner (1965). He begins with a quotation of Rumpf which states that "Everybody practising comminution knows that this always was and still is today an empirical discipline".

A quotation from the rules of VDZ (Union of German Cement Works, FRG) states that: "The filling rate and the charge composition must be determined by experiment".

For dry grinding, there does not and what is more there cannot exist a wear function of general validity. For wet grinding where the wear by corrosion is overwhelming—and surpasses 5–7 times that in dry grinding—the surface effect is unambiguous therefore the exponent 3.84 of the equilibrium ball size distribution (Bond, 1958), or theoretically 4 (Paulsen 1964), as seen in formula (13.11) is valid. With dry grinding, its value can be between 2 and 3.

To conserve a well proved ball size composition, the replacing of the worn mass by single sized balls does not conform with the requirements.

For coarse grinding a mixture of different sized balls is convenient. This must be composed, according to Slegten, of the different sized balls in equal number; according to Bond, in conformity with equilibrium distribution.

For fine grinding the ball size distribution must be in conformity with surface increase. In open circuit a small porosity is suitable while in closed circuit operation a large throughput, i.e. larger free space between the balls is required. In closed circuit a less carefully composed ball charge can be tolerated.

Symbols in Chapter 13

D	mill shell diameter, m ev. feet (or as below)
L	mill shell length, m
d	grinding body size, mm ev. inches
x	particle size, mm or micron
B	grindability index, cm^2/J
Wi	work index, kWh/sht
K, c, g, k	constants
δ	density, t/m^3
Cs	percentage of critical rpm
l	distance from mill inlet, m
G	grinding body mass or weight, kg (or as below)
S	grinding body surface, m^2
t	grinding time, hours
D	percentage passing in grinding body size distribution (in Paulsen's demonstration)

G	relative undersize mass in the charge (again in Paulsen's demonstration)
z	number of replaced balls
Q	mill charge, t
N	energy demand, kW
M	liner wear, t
ε	filling rate, percentage
α, h	as in Fig. 13/7.

CHAPTER 14

PROCESSES IN ROLLER MILLS

Roller mills resulted in the systematical development of the classical chilean or pan mills. As is well known, in chilean and pan mills only the centre plane has a slide-free rolling movement, even this slide friction work constitutes the main part of its energy consumption. The widespread types of high capacity roller mills avoid this energy waste by special roller shapes: Loesche mills have conical, Pfeiffer mills sphere segment shaped and Polysius mills vertically divided rollers. Pressure upon the rotating bowl is increased by springs or by hydraulics. In the latter arrangement, raising of the rollers during idling enables friction to be entirely eliminated.

In their movement the rollers exert a heavy pressure upon the material bed and bearing in mind the metre-sized rollers it is clear that agglomeration is unavoidable (see formulae (4.4) and (4.5)). On the other hand roller mills are always air swept; compared with ball mills the retention time is smaller by an order of magnitude and small particles are taken away by the air stream. In consequence roller mills are instruments of grinding to moderate fineness; usually a 12–15% residue is not surpassed on the 0.09 mm sieve. Nowadays experiments are in progress on fine grinding (cement grinding) in roller mills (Schauer 1977).

From the standpoint of energy consumption, compared with ball mills roller mills sometimes offer considerable advantages: energy waste by ball friction and ball collision does not take place. Besides rolling friction, a considerable amount of energy is consumed by the ventilators. In large spaced roller mills it is, however, smaller than with air swept ball mills, which is an important factor in dry grinding operations.

Experiments executed by the firm Allis Chalmers, USA, with closed circuit ball mills and Pfeiffer type roller mills enabled Klovers (1973) to demonstrate energy savings in the range of 15–25%—to the latter's advantage. In roller mill operations ventilation requires about 35–40% of the whole energy consumption. Experiences in the Swiss Holderbank cement factories prove a saving of about 25% in maintenance costs (wear repair) related to those of ball mills (Schildknecht 1978).

Further advantages are that the feed size in large roller mills can reach or surpass 100 mm, so secondary crushing can be eliminated. The changing of worn rollers requires only a small part of the time demand of the change for ball charge, i.e. the availability index is more favourable. And one should not forget the relatively noise-free operation.

CHAPTER 15

PROCESSES IN IMPACT MILLS

The behaviour of brittle materials on crushing following the impact by single solid bodies was investigated by Okuda (1971). He stated that

- — the shape of product particles changes with the impact velocity: low energy impact produces elongated, higher energy impact stumpy particles,
- — there exists a limit value of the reduction ratio which itself is dependent on impact velocity,
- — there exists a grinding velocity furnishing an optimal efficiency,
- — the breakage takes place in three steps: initial breakage; further impact breakage of the produced particles; and thirdly, the mutual crumbling of the breakage particles.

Processes of impact mills can best be understood by the laws of collision. The significant difference between ball mills and impact mills is to be found in the mode of energy transmission: in ball mills the kinetic energy performing the breakage is carried by the grinding bodies whereas in impact mills by the particle to be comminuted.

From the principles of mechanics it is well known that the laws of collision can be derived from the constancy of impulses.

If two bodies of masses m_1 and m_2 collide and their velocities before and ollowing collision are respectively u_1 and u_2 and v_1 and v_2 then

$$m_1 v_1 - m_1 u_1 = \int\limits_0^\tau P \, dt$$

and

$$m_2 v_2 - m_2 u_2 = - \int\limits_0^\tau P \, dt$$

that is, an impulse gain of m_1 is equal to an impulse loss of m_2

$$m_1(u_1 - v_1) = m_2(v_2 - u_2). \tag{15.1}$$

During collision, the colliding bodies suffer deformation. In the first phase they are compressed and in the case of ideal elastic bodies they regain their original form without loss of energy; with ideal rigid bodies this energy will be lost. In our case, i.e. with brittle materials, the process is quite the same, the energy of collision will be so to say totally lost, this energy loss is the source of crushing. If the energy demand of breakage is surpassed, crushing takes place.

This energy can be calculated as follows: At the end of the first phase of collision both bodies have the same velocity, viz.

$$u = \frac{m_1 u_1 + m_2 u_2}{m_1 + m_2} \tag{15.2}$$

The kinetic energy of both bodies is

$$W_u = \frac{1}{2}(m_1 + m_2)u^2$$

and the energy before collision is

$$W_1 = \frac{1}{2}(m_1 u_1^2 + m_2 u_2^2)$$

thus the energy loss, in our case the crushing energy, is given by

$$W = W_1 - W_u = \frac{m_1 m_2}{m_1 + m_2}\frac{(u_1 - u_2)^2}{2} \tag{15.3}$$

If crumbling takes place, the velocity according to formula (15.2) represents the velocity of the common mass centre of the disintegrated particles.

For rotating impact mills we shall simplify the problem to perpendicular collision, where the rotating blade has an overwhelming mass $(m_1 = \infty)$ and the particle to be ground is in the resting state $(v_2 = 0)$. So

$$W = \frac{m_2 v_1^2}{2} \tag{15.4}$$

As seen above, the relative velocity of the particle determines the crushing energy. Taking into account the classical volume theory of comminution based on formulae (1.2a) and (15.4), the velocity to perform crushing can be calculated

$$\frac{mv^2}{2} = \frac{m}{\delta}\frac{\sigma^2}{2E}$$

where E is Young's modulus, σ the crushing strength and δ the density of the particle, and

$$v = \frac{\sigma}{\sqrt{\delta E}}. \tag{15.5}$$

Introducing the common formula of sound velocity in the material $a = \sqrt{E/\delta}$ we get

$$v = a \frac{\sigma}{E}. \tag{15.5a}$$

This is a very interesting result: the velocity capable of performing crushing is seemingly independent of the size of the particle. Under actual conditions this does not hold: the strength increases with decreasing dimensions owing to the decreasing number of defect locations. This gains importance in the sphere of very fine grinding (see Chapter 16).

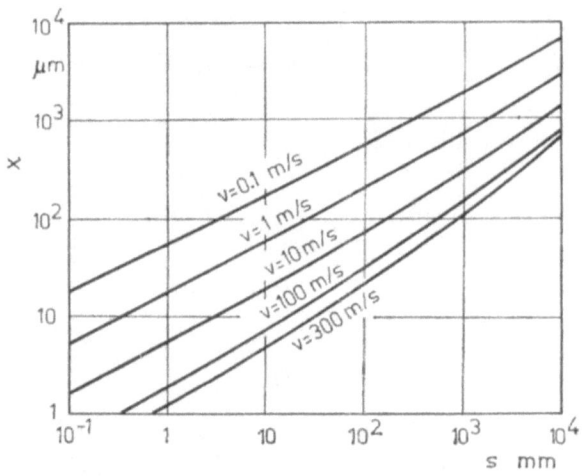

Fig. 15/1. s length of trajectory in mm of spheres of size x; density 1 g/cm³, v initia velocity, according to Rumpf (1959)

Taking, for example (in common and in SI units) $\sigma = 2000$ kgf/cm² $= 2 \times 10^8$ Pa, $\delta = 3 \times 10^3$ kg/m³ and $E = 600\,000$ kgf/cm² $= 6 \times 10^{10}$ Pa we get $v = 15$ m/s in agreement with usual rotating speed (10–50 m/s) of impact crushers.

In the powerwork practice widely used shock wheel (or ventilator) coalmills the circumferential rotating speed is about 80 m/s to attain the grinding fineness of about 5–30% residue on the 0.09 mm sieve, in harmony with the firing properties of the type of coal.

An outstanding analysis of the processes of impact grinding can be found in a paper by Rumpf (1959). We give here some extracts from his elaboration:

For collision velocity he recommends in general terms

$$\frac{\sigma}{E} = f\left(\frac{\delta v^2}{E}\right) \tag{15.6}$$

and in particular

$$\frac{\sigma}{E} = 0.328 \left(\frac{v}{a}\right)^{0.4} \tag{15.6a}$$

that is, he attaches lesser importance to the shock speed.

According to formulae (15.5) and (15.6) the mutual collision of accelerated particles as well as their impact on the circular lining will determine the performance of grinding. Rumpf investigated therefore the velocity decrease in consequence of air resistance and mutual collisions.

In consequence of the air resistance along trajectory s, the starting velocity v will be eleminated. The values of s for particles of size x and density 1 g/cm³, at an air temperature of 20 °C are presented in Fig. 15/1.

The free path limited by mutual collisions was calculated by Rumpf based on Maxwell's speed distribution law

$$\lambda = \frac{x}{10(1 - \varepsilon)} \tag{15.7}$$

where ε is the porosity or $1 - \varepsilon$ is the ratio of the space occupied by the particles. With advancing grinding x and λ will simultaneously decrease.

Both limiting factors are presented in Fig. 15/2. It is clear that in most cases λ will have the determining role. So, for example, for $x = 100$ micron, λ will be between 1 and 10 mm; below 10 micron, collisions will be very frequent, their effect will be decisive and overshadow that of rotating plates.

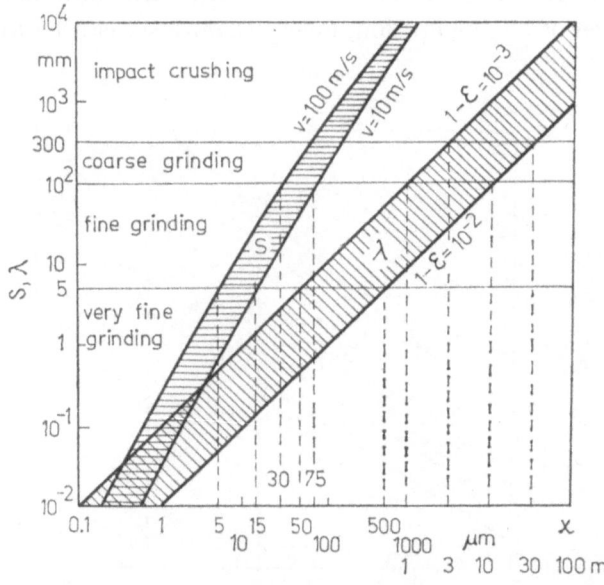

Fig. 15/2. Trajectory limited by air resistance (s) and by mutual collisions (λ) in 20 °C immovable air, x particle size, density 1 g/cm³ (Rumpf)

Rumpf's paper concludes by dealing with considerations relating to energy demand. He supposed that the first collision results only in fatiguing the material and that breakage occurs only after further collisions. Collisions of unsufficient energy as well as these of already ground particles are wasted—these latter must be removed from the mill space by accurate methods, i.e. by closed circuit. All these conditions result in an energy demand proportional to the accelerating velocity, as summarized in Fig. 15/3.

A practical formula for the energy demand of rotating impact mills (shock wheel or ventilator mills) was proposed by Boross (1969)

$$N = \frac{u}{\eta_m} (\delta_g V_g v_g + m_p v_p) \tag{15.8}$$

which involves the work necessitated in accelerating the gas and the powder. Here u is the circumferential velocity of the rotor, η_m the mechanical efficiency, δ_g the density, V_g the volume per second, v_g the velocity of the gas leaving the mill, m_p the mass per second, and v_p the velocity of the powder, all measured in the outlet in m–kg–s system.

Planiol (1962) proposed that the wheel be rotated in a vacuum to eliminate the effect of air resistance. This equipment is still being developed. For these vacuum centrifugal mills, the first term in formula (15.8) becomes equal to zero.

The next problem to be analysed is that of agglomeration. Here the conditions are very favourable in comparison with ball or roller mills. As seen above, the carrier of breakage energy is the particle itself, with suitable collision velocity the main source of agglomeration, energy overdose, can be avoided. And one

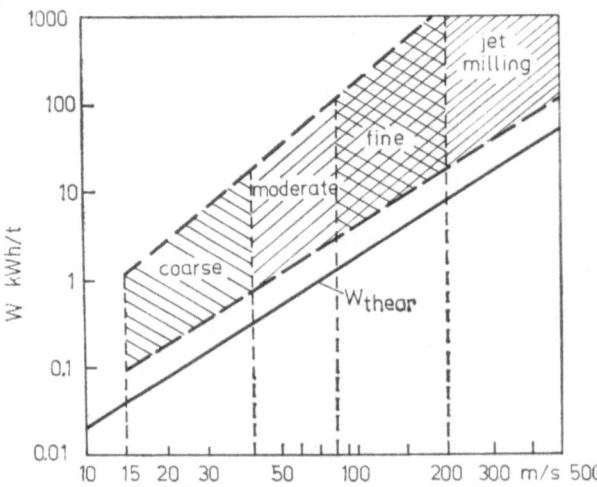

Fig. 15/3. Specific energy demand against collision velocity (Rumpf)

other favourable condition: there is no material bed, the particles are loosely dispersed in the grinding space.

A numerical example will be very instructive: In a ball mill with balls of 30 mm or 0.11 kg mass and a cataracting height of 2.5 m, the shock energy is $2.5 \times 0.11 = 0.275$ mkgf or 2.7 J.

In an impact mill a particle of size 1 mm, $\delta = 3$ g/cm³ density, mass $1.57 \times \times 10^{-6}$ kg and a collision velocity of 80 m/s will have the collision energy 5.12×10^{-4} mkgf or 5×10^{-3} J. The energy dose in a ball mill surpasses, by more than two orders of magnitude, that in the impact mill. And for a particle size of 0.1 mm this energy dose decreases by a further three orders of magnitude.

As for maintenance, because of the great collision speed, a considerable amount of wear can be expected. Wear resisting alloys are somewhat different from those proved good for ball mills: brittle paddle blades will crumble and the possibility for repair by welding is essential. Single baldes or even the whole rotating wheel can be changed in a few hours, in larger machines this is done with the aid of hydraulic devices. The mass of the mill and especially that of wearing parts is small compared with ball or roller mills. Because of this the availability factor is favourable.

As for capacity: nowadays the biggest mills in powerworks have a driving power of about 1000 kW, this is far behind the requirements in up-to-date cement and ore dressing plants.

Symbols in Chapter 15

m	mass of colliding bodies, kg
u and v	collision velocity, m/s
$\int_0^\tau P\,dt$	impulse, mkg/s
W	work or energy, J
δ	density, t/m³ or kg/m³
σ	crushing strength kgf/cm² or Pa
E	Young's modulus, kgf/cm² or Pa
a	sound velocity, m/s
s and λ	free flying path, mm or microns
x	particle size, mm or microns
ε	porosity
η_m	mechanical efficiency
V_g	gas volume throughput, m³/s
v_g	gas velocity in outlet, m/s
m_p	powder output, kg/s
v_p	powder outlet velocity, m/s

CHAPTER 16

VERY FINE GRINDING

Using the somewhat arbitrary definition, we call "very fine grinding" the process from which the product size is below 40 microns but requirements sometimes reach the minus 3 microns range.

As expounded in detail in the preceding chapters, fineness increase is hindered by aggregation, and agglomeration and is a consequence of an excessive energy supply. On the other hand the strength increases with decreasing dimensions owing to the decreasing number of defect locations; in order to reach breakage, a larger energy concentration is needed.

To solve these seemingly contradictory requirements the energy must be supplied in a great number of individually small energy doses. The other solution is the closed circuit operation with special classifier constructions (see Chapter 18). We shall, of course, meet the combination of both methods.

Vibration mills represent the one kind of very fine grinding. As generally known, the cylindrical or trough-shaped vessel is 70—90% filled with steel balls of about 10 mm diameter—or if iron contamination is not allowed, with ceramic balls; it is suspended on or supported by springs. By rotation of out-of-balance weights a vibrating system is set up. The grinding vessel will have a circular or elliptical path with an acceleration surpassing 3–10 times the gravitational acceleration, the amplitude is not greater than some millimetres. The mostly preground feed occupies the space between the grinding bodies and is exposed to a great number of shocks with a small amount of energy.

According to literature data, with accurately preground feed a product of 1–10 microns was obtained. The customary product size is about 20–40 microns.

The mechanics of vibrating mills was first investigated by Bachmann (1940). According to his analysis advantageous performance can be attained in the state of "statistical resonance" when the period of vibration is equal to the trajectory time of grinding bodies.

The main movement of the grinding bodies and the feed is composed of microtrajectories. Because of air resistance the feed will successively slown down thus collisions take place resulting in breakage by shock and rubbing.

Besides this main movement the whole mass will slowly circulate in the opposite direction to the grinding vessel oscillation and the balls will have their own rotation too as seen with ball mills (Fig. 12/5).

Effectivity is influenced by frequency, amplitude, filling rate and properties of the feed. The effect of filling rate was investigated by Iwata et al. (1974).

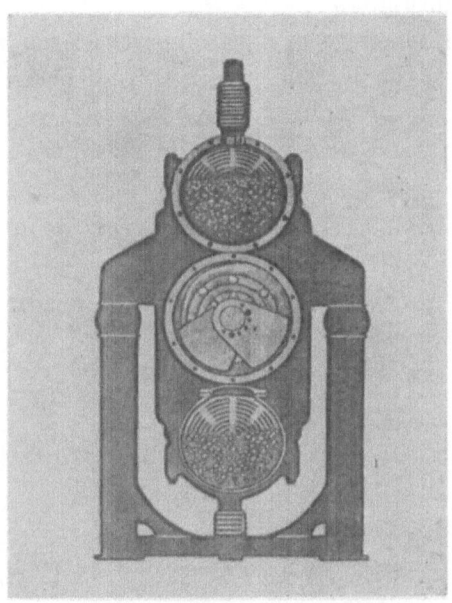

Fig. 16/1. "Palla" mill (KHD)

According to their findings, the free space above the charge level will determine optimal conditions: a possibly high filling rate not impeding, however, the development of free trajectories of the upper ball row must be ensured.

According to Rose (1962) a large speed of vibration is required. The adaption of ball size to strength and size is essential. Increased speed of vibration causes, however, considerable bearing friction and stress of the foundation. This latter can be avoided by multivessel, mass-balanced arrangements (Fig. 16/1, KHD).

In continuous operation, two stage arrangements (grinding body charge in the first stage, e.g. 20 mm balls; in the second, 10 mm balls) in a closed circuit are frequently applied.

The energy demand of vibration mills is composed of: energy dissipated in bearing friction, by wind resistance of the out-of-balance weights, by metal hysteresis of the supporting springs, and the useful energy transferred to within the mill space.

Energy consumption related to the mass of charge remains slightly behind that of ball mills (this latter is equal to $C\sqrt{D}$ according to formula 12.5). Because

109

of the high filling rate (small free space above the charge), the process is sensitive to changes in material hold up, in closed circuit choking can occur.

Special impact mills can verify the accurate energy dosage without contact of particles in a material bed thus ensuring elimination of agglomeration.

As seen in the previous chapter, to attain high grinding fineness the collision velocity must be increased naturally with increase of specific energy consumption as summarized in Fig. 15/3.

The air stream influencing the path of particles has an important role by promoting collisions of the particles with the rotor blades and with each other.

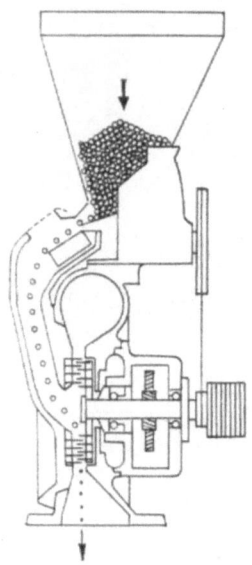

Fig. 16/2. "Contraplex" mill (Alpine AG.)

The other role of the air stream is the cooling of the mill space which thus reduces the energy level responsible for agglomeration.

A large assortment of such constructions is at our disposal. Without the slightest intention of being biased in favour of any one of these constructions, we present in Fig. 16/2 the "Contraplex" mill of Alpine AG. (FRG) and in Fig. 16/3 the "Super micron" mill of Hosokawa Co. (Japan).

In the first of these, two wheels are rotating with different speeds mostly in opposite directions thus doubling the shock velocity. The biggest unit has a driving power of 150 kW, a fineness below 30 microns, its output in function of fineness and grindability can reach some tons per hour.

The second is a two step pulverizer adjustable to grinding finenesses dependent upon grindability in the range of 140 to 2 microns. Apart from the very fine grinding a selectivity is ensured by removing contaminating coarse par-

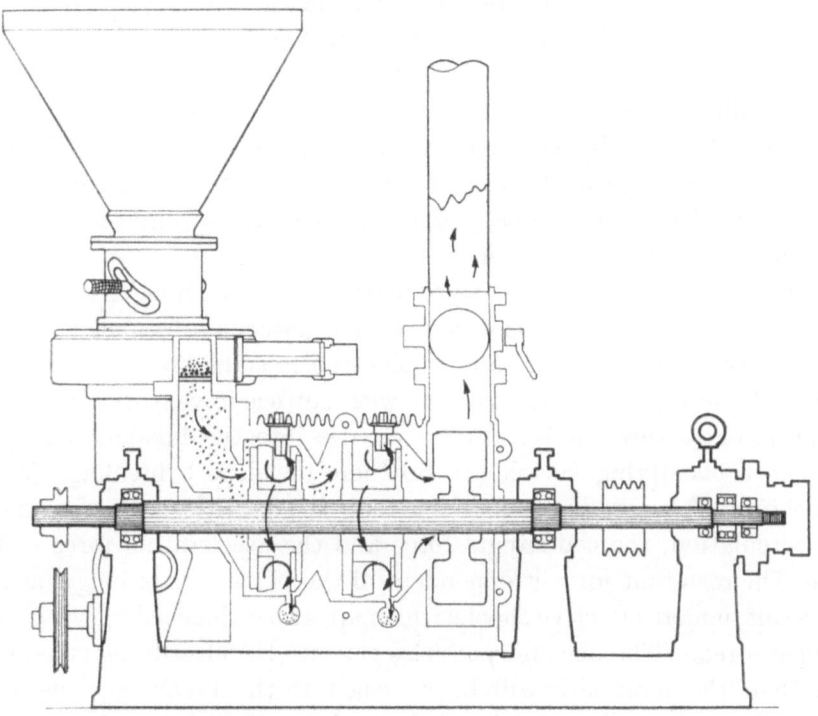

Fig. 16/3. "Super micron" mill (Hosokawa Co.)

ticles of worse grindability. As much as 40% of the feed can be extracted by means of this optional up-grading device. There are models of 5 to 55 kW driving power.

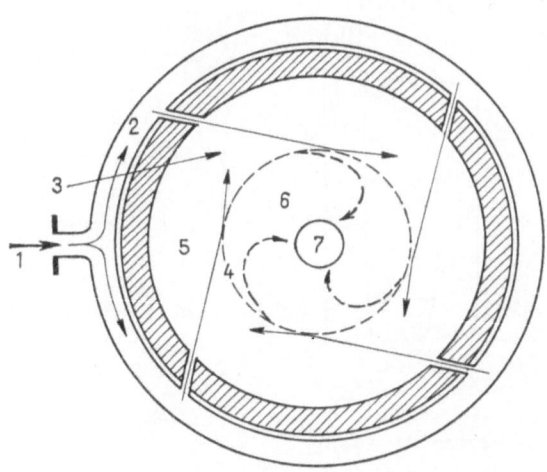

Fig. 16/4. Principle of jet grinding. For explanation, see text

111

Growing air resistance and tensive stress in the rotor material constitutes an upper limit of rotating speed, according to Fig. 15/3 it cannot exceed about 200 m/s.

Bigger collision velocities can be attained in compressed air or in a high pressure jet of steam. So we arrive to *jet mills*, devices of finest grinding with product sizes below 5, sometimes below 1 μm. There are vertical and horizontal arrangements. In another type, nozzles are located on opposite sides giving rise to countercurrent jet stream.

The working of jet mills can be explained if we examine Fig. 16/4, which illustrate the scheme of a horizontal mill. The gaseous substance (1) is led into the high pressure chamber (2) and to the feed injector (3). From the pressure chamber the gas gets through nozzles with sufficient velocity, in the case of extreme fineness through Laval-type nozzles into the breakage chamber (5) producing a classifying vortex (4). So there develops a breakage (5), and a classification (6) zone. Three forces act on the particles: the dragging force of gas circulation, the centrifugal force and the accelerating force of the gas stream. The resultant force is dependent of the particle size; bigger size differences result in more effective shocks which are also influenced by the turbulence of the gas stream. The once hit particles get into the classifying zone, particles bigger than the limit size will be returned to the breakage zone and once again accelerated, on the other hand the cyclone effect causes the fine particles to be removed downwards from the mill space, whereas the gas leaves through the upper outlet (7). A considerable recirculation represents the real charge of the mill.

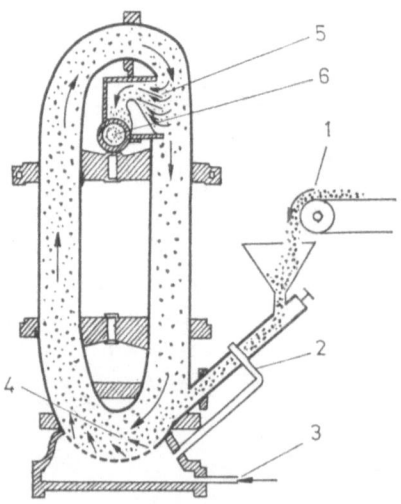

Fig. 16/5. "Jet-O-Mizer" (Fluid Energy Int. Sales Co.)
1 feed; 2 injector; 3 compressed air; 4 breakage chamber; 5 classifying zone; 6 outlet

Neglecting the very extensive literature we refer to a paper by Mori (1962) summarizing the processes in jet mills. The pulverizing takes place primarily in the jet stream at the vicinity of the nozzle. Body breakage and surface grinding take place simultaneously. The grinding rate is proportional to the cross-sectional area of the nozzle and to the nozzle pressure. The recycling frequency of particles (analogous to circuit number in closed circuit grinding) is reciprocally proportional to the feed rate and its grindability. The capacity increases with the compressor power to the 2–3.3 power and the value of this exponent can be approximated as 2.5.

This latter statement is of great importance because nowadays the largest capacities do not exceed some tons per hour.

Up to date jet mills work at supersonic speeds. In a paper by Korda (1961), such conditions are analysed. The physical characteristics of different gaseous substances and the effects of convergent–divergent nozzles are dealt with in particular. Suitable Laval-type nozzles and saturated steam enable 1250 m/s to be achieved.

Jet milling produces a narrow particle size range – which is a requirement to avoid agglomeration (this narrow particle size being an advantageous property of cement, see Chapter 19) (Deshko et al. 1973).

As for the vertical type jet mill in Fig. 16/5 the well known Jet-O-Mizer is presented.

Figure 16/6 illustrates the counter current jet mill (Rumpf 1960), yielding a possibility for somewhat coarser grinding. Details on this are given, for example, in a paper by Akunow and Deschko (1978) where investigations for cement grinding were reported.

And here a short reference to the wet process colloid or rubbing mills with high rotation speed. In consequence of a high speed stream through narrow slits between plane, cylindrical or conical surfaces, sheet stress will produce a very fine homogeneous paste.

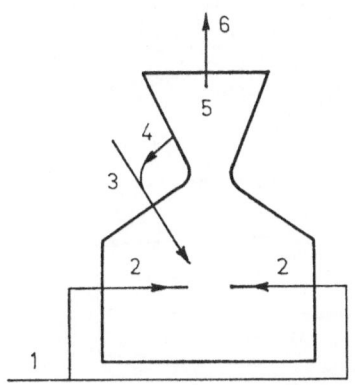

Fig. 16/6. Counter current jet mill
1 gas inlet; 2 nozzles; 3 feed; 4 tailings; 5 separator; 6 fine product

CHAPTER 17

WET GRINDING AND DRYING-GRINDING
OPERATIONS

By wet grinding the nature of the pulp hinders the direct contact of particles, agglomeration becomes likely only in a very late phase of the process following the thickening of the pulp. The moisture content of the slurry has a determining influence upon efficiency, an optimal value can always be found. In the case of cement raw meal grinding as in Fig. 17/1, sieve residues $R(0.09)$ and $R(0.063)$ are presented as a function of the water content in the slurry related to one kilogram of dry substance. It can be realized that in this case optimal conditions

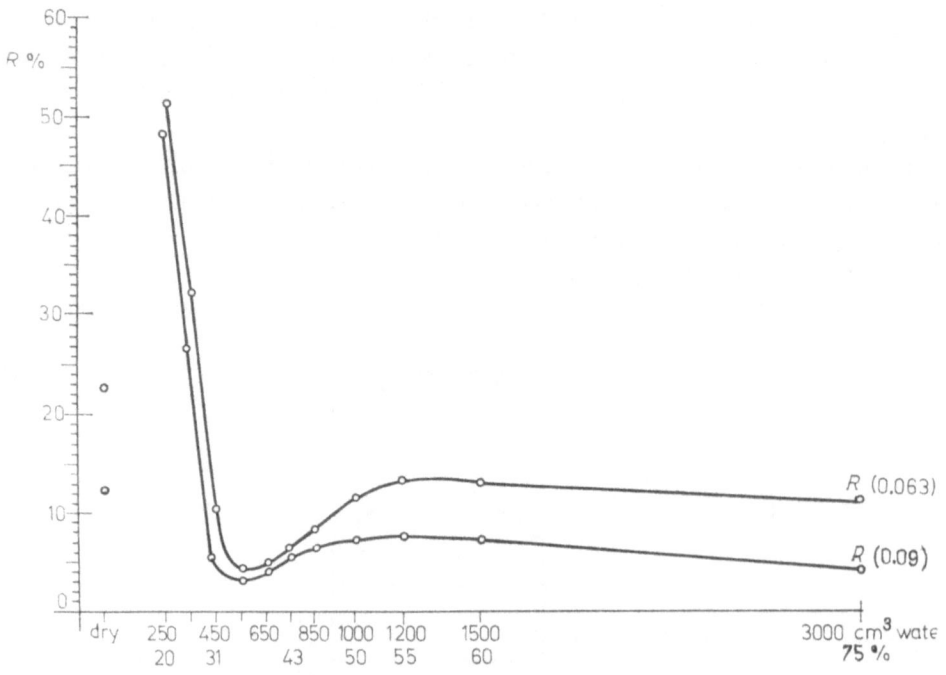

Fig. 17/1. Wet grinding of cement raw mix. Sieve residue against water content in the slurry by identical energy consumption

114

can be reached when the mass of water is about half of the dry substance or a third of the slurry. So the grinding efficiency of wet grinding can surpass that of dry grinding by some 20–25%. But the lining and grinding media wear may well reach a multiple which can be 5 to 7 times the value. Several production technologies exclude wet grinding (e.g. cement grinding); for other technologies it is self-evident (e.g. preparation for flotation); in other cases economical considerations are responsible for the choice of the appropriate method.

An important advantage of the wet process is that there is no problem with dust collection or heat generation.

In wet ore grinding operations the filling rate sometimes reaches or even surpasses 50%. On the other hand a moisture content surpassing about 1–2% (influenced by properties of feed and function of equipment) spoils the efficiency of dry grinding and a stream of hot gases through the mill space must be used. In basic terms, all airswept mills are suitable for this purpose. In ball mills sometimes a grinding chamber with lifting blades and without grinding media is built in. The sometimes large-sized feed (up to 80 mm) needs the use of an impact precrusher through which also the drying hot gases flow. In other cases the hot gases do not get through a ball mill which has a high air resistance, in which case the drying takes place in the precrusher and separator, this latter must of course be connected to a dust collector. But it is necessary to note that a water content of some parts in a thousand is favourable because of its tensio-active properties. This also promotes the effectivity of electrofilters. Some

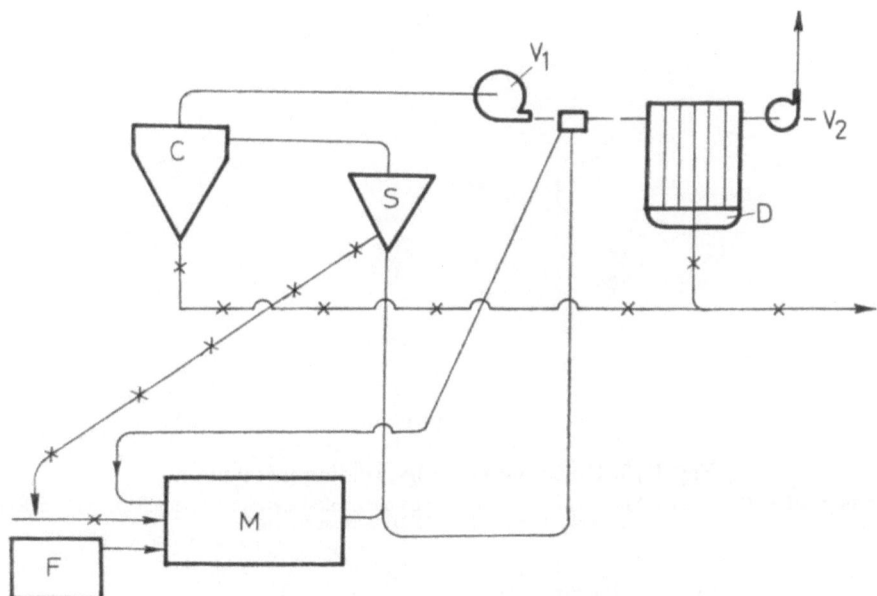

Fig. 17/2. Air swept drying-grinding installation

M mill; *S* separator; *C* cyclone; *D* dust collector; V_1 circulating; V_2 evacuating ventilator; *F* furnace

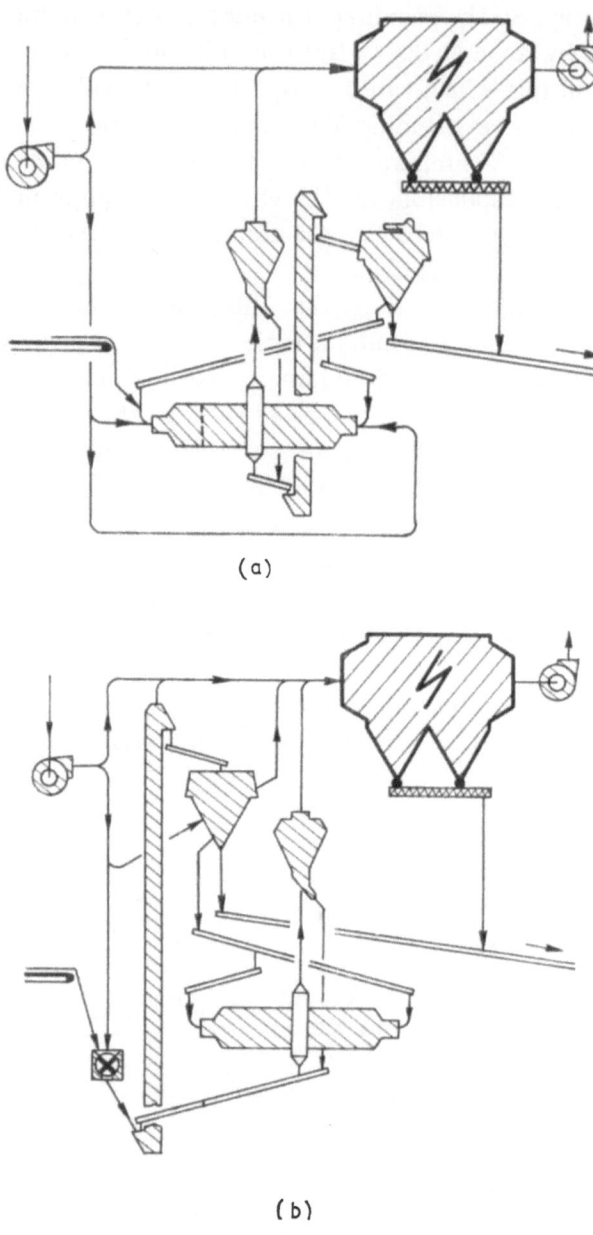

(a)

(b)

Fig. 17/3. Up-to-date drying-grinding installations
(a) "double rotator" (Polysius); (b) hammermill precrusher and mechanical elevator (Hischmann); (c) "tandem mill" — precrusher and airlift (KHD) (Fasbender 1971)

twenty years ago air swept ball mills with air recirculation were widespread but now they are utilized very rarely because of the high energy demand of the ventilators (Fig. 17/2).

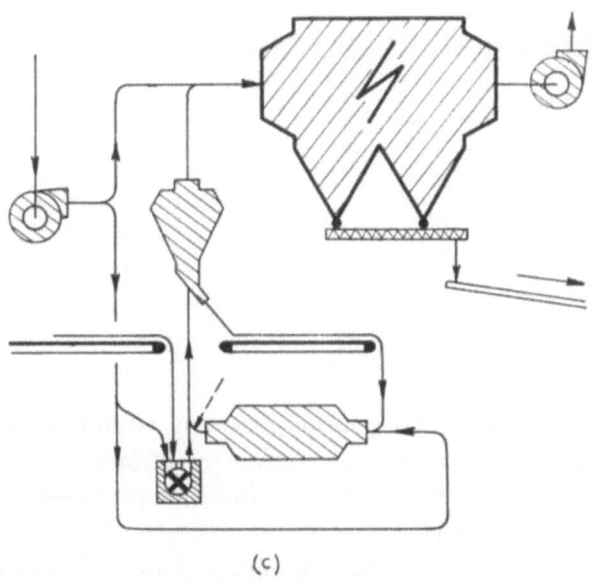

(c)

Fig. 17/3. (cont.)

Three up-to-date types of ball mill drying-grinding installations are illustrated in Fig. 17/3 (Fasbender 1971).

The previously discussed aerofall mills, roller mills and impact mills are of
course very convenient for simultaneous drying and grinding.

Some characteristic data related to the drying-grinding of cement raw meal
according to a paper by Erni (1971) are as follows:

Gas temperature	350°	900°
Single chamber ball mill with mechanical air sep- arator	0–3	3–8% moisture content
Double rotator	3–6	6–12% moisture content
Ball mill with heated impact crusher	3–8	8–12% moisture content
Airswept ball mill	5–7	7–12% moisture content
Airswept ball mill with heated impact crusher	7–12	12–15% moisture content
Roller mill	0–12	12–15% moisture content
Aerofall mill	0–12	12–22% moisture content

CHAPTER 18

PROCESSES IN CLASSIFYING EQUIPMENT

Grinding fineness in the *dry process* in closed circuit operations is ensured by air separators (air classifiers, air sizers). In conformity with the preceding, we shall confine our discussions to the internal processes in separators and ignore the construction details.

Gravitational separators are the classifying element in airswept mills. With roller- and impact mills they are directly built together, with ball mills they

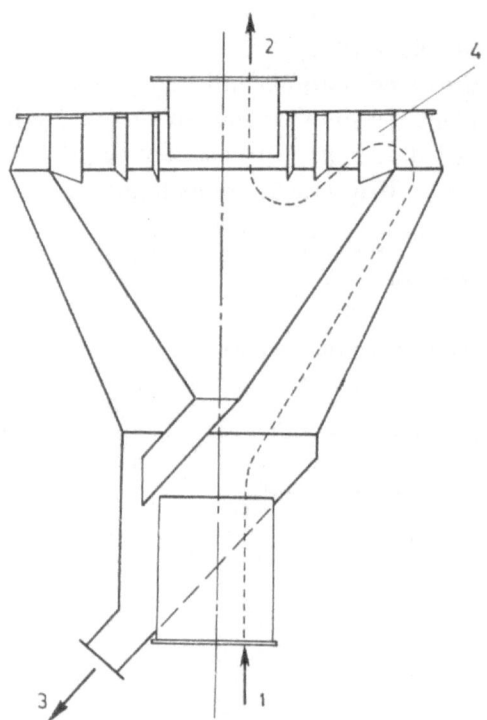

Fig. 18/1. Gravitational separator
1 air and pulver inlet; 2 fine fraction and air outlet; 3 coarse fraction outlet; 4 vane

are attached by short pipe sections (see e.g. Fig. 17/2 and 17/3c). In principle this is a coaxial double cone connected by an air vane ring (Fig. 18/1). Below, the air and material inlet is to be found. In the upward stream the air velocity decreases, large sized particles will gradually drop out. At the upper part the adjustable vane induces centrifugal forces and a cyclone effect to separate somewhat smaller particles. The fine fraction leaves the separator upwards, together with the airstream. An outer ventilator and dust collector completes the installation.

With a constant airstream the separating size can be regulated by opening or closing the air vanes. With an increasing air stream in the outer cone the separating size increases, it decreases in the vane and cyclone with the result that no decisive change will occur but the quality of separation deteriorates. These devices allow a separating fineness characterized by $R(0.09) = 12-15\%$ to be attained.

Mechanical or *dispersion* separators are composed of the outer and inner cylindrical-conical drum (Fig. 18/2). The material passes through an upper or lateral inlet to the rotating distribution plate. The same vertical shaft rotates both the main and the auxiliary fans. The main ventilator sets up an upward stream which elevates the fine particles throwing them into the outer fines cone whereas the tailings fall downwards in the inner cone. The circulating air

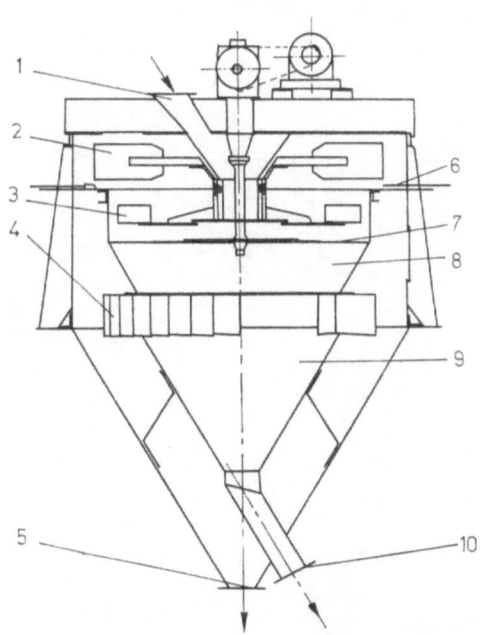

Fig. 18/2. Dispersion separator (Sturtevant)

1 material inlet; 2 main ventilator; 3 auxiliary ventilator; 4 separating vane; 5 fine fraction outlet; 6 control valve; 7 distribution plate; 8 separating space; 9 tailings cone; 10 tailings outlet

returns through the air vanes. The separating size is decreased, the separating effect is improved by the effect of an auxiliary ventilator which develops a spiral stream and the centrifugal force pushes even finer particles to the outer drum plate where they lose their velocity and get to the tailings discharge.

A change of rotating speed affects both ventilators in an adverse sense. In some up-to-date constructions therefore, they are independently driven by two motors and a hollow shaft, the auxiliary ventilator is speed controlled.

The velocity of the air stream can be regulated by control valves too.

Performance is also affected by the load of material. With a big load, growing concentration increases the air resistance, air velocity decreases, and the separating size decreases too. But in consequence of growing concentration, particles of different size—and accordingly different velocity—will collide, larger and smaller particles get mixed, the effectivity of separation deteriorates. This means that a good separation can be achieved by ample separator dimensions.

The fine fraction is separated from the circulating air stream by the return air vanes; in terms of quality, it is far from a perfect dust collector. A proportion of very fine particles will therefore return to the separating zone and ultimately get to the tailings. This effect was mentioned on page 68.

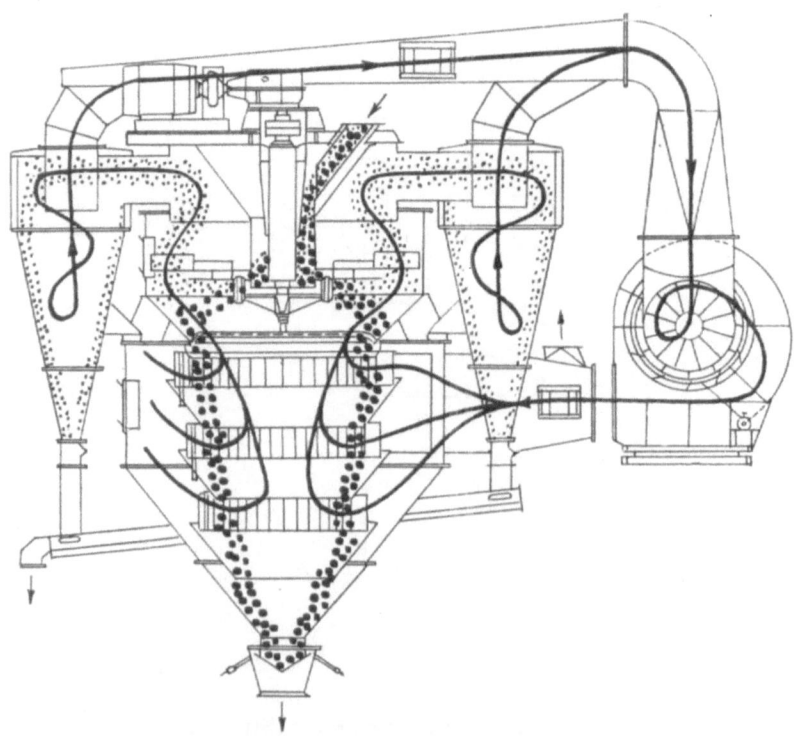

Fig. 18/3. Dispersion separator with outside cyclones and ventilator (WEDAG)

The separating effectivity can be improved by connecting two separators (as dealt with in Chapter 10).

In up to date separator constructions (first introduced by Wedag, FRG), air circulation is produced by a ventilator outside the sizing space and the fine fraction is collected by cyclones placed around the separator cone (Fig. 18/3). In such devices the ventilator is not damaged by dust, the cyclones can separate fine particles much better than the return vanes. The output related to cross section of the sizing space surpasses by some three times that of classical separators (the section of cyclones excluded).

The lower size limit of these mechanical air separators is about 45 microns.

An important problem is the mutual effect of mill and separator in the case of grindability changes. With deteriorating grindability the mass of tailings and with it the throughput of mill will increase. This again, as seen above, decreases the separating size. In consequence, the mass of tailings and throughput suffers a further increase. To avoid choking of the mill the feed must be reduced. But, bearing in mind the retention time of the installation, a fluctuation will develop; to prevent this inconvenience acoustic regulation can be applied.

Air sizers are sometimes utilized for heating (drying-grinding operations, see Chapter 17), or even for cooling the ground product. For this purpose hot/cold air or gases are introduced into the sizing space. This mass of gas must of course be let out—passing through a dust collector.

Extreme fineness can be attained by the "microplex" separator (Alpine AG., FRG) or similar devices. Separation takes place here in a centrifugal field,

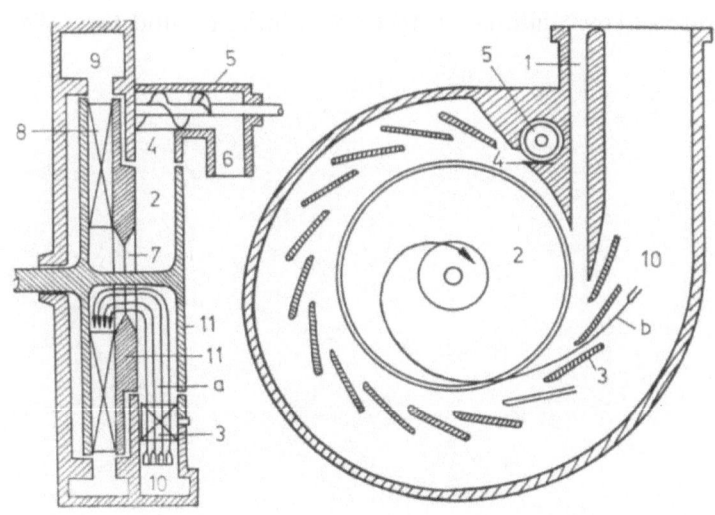

Fig. 18/4. "Microplex" separator (Alpine AG)

1 material inlet; 2 separating space; 3 adjustable vane; 4 separating wedge; 5 screw conveyor; 6 tailings outlet; 7 throughflow slot; 8 ventilator; 9 spiral house; 10 air inlet; 11 rotating border plate

the bordering plate of the sizing space rotates with the ventilator thus neutralizing the wall friction (Fig. 18/4). The size limit changes with output, according to data by the manufacturer: with 132 mm sizing space diameter it is 2–15 microns and 50–300 kg/h; with 800 mm dia 10–40 microns and 1200–1600 kg/h.

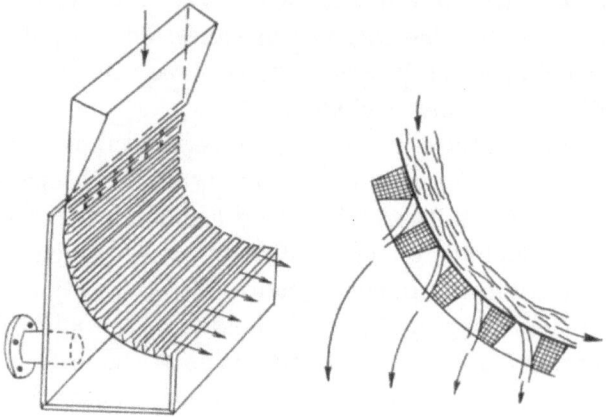

Fig. 18/5. DSM arc screen

In the *wet process* rake classifiers have the separating size of some hundreds of microns, hydrocyclones are one order of magnitude smaller. It is important to note that for a tolerable separating efficiency, the moisture content in the slurry must be at least 70% —which is not acceptable for some manufacturing processes.

The DSM arc screens (fanlike grates) are gaining ground too (Fig. 18/5).

CHAPTER 19

SATISFYING TECHNOLOGICAL REQUIREMENTS

Grinding is always part of a production technology and the requirements set against the milling product are determined by the technological process. These requirements prescribe mostly an upper limit of sieve residue but sometimes a rate of some definite size fractions is required. It is, of course, a difficult matter to satisfy this latter requirement because it is in contrast with the natural particle size distribution and can only be approached by linking the classifying procedure.

In the following, grinding problems of some production technologies will be dealt with; the emphasis is mainly on the production technologies of cement grinding since these are instructive for other technologies too.

19.1 CEMENT GRINDING

For cement grinding, solely ball mills are used. A comprehensive description of cement grinding equipment and methods was presented by Sillem (1972 and 1977) for the Verein Deutscher Zementwerke (VDZ) Congresses. More detailed data are collected in Duda's "Cement Data Book" (1976–1977). The strength and the hardening rates of cement are determined for a constant chemical–mineralogical composition by the particle size distribution. Several attempts have been made to define the optimum value for this but no generally accepted result has been obtained. The author's own experiments and theoretical considerations as well as the study of a vast amount of literature have, however, led to the conclusion that the maximum possible percentage of particle size fraction between 3 and 30 microns is essential: too small particles get hydrated too early before the preparation of concrete, and moreover are usually poor in clinker minerals; too large particles, on the other hand, are hydrated only superficially and not throughout their entire volume. Particle size limits of 3 and 30 microns are, obviously, only tentative representing an order of magnitude; by defining the lower limit as 2 or 5 microns or the upper limit as

20 or 50 microns will not affect the results below. Our own investigations claim as a value, 70% of this fraction in order to get a super duty cement, whereas for moderate hardening 45–50% is sufficient.

Hydration takes place at the surface of the particles which means that a high specific surface is required for rapid early hardening. The bulk of the surface is given by the smallest particles, say 0–3 microns. Therefore this fraction will also be taken into consideration in the following discussions.

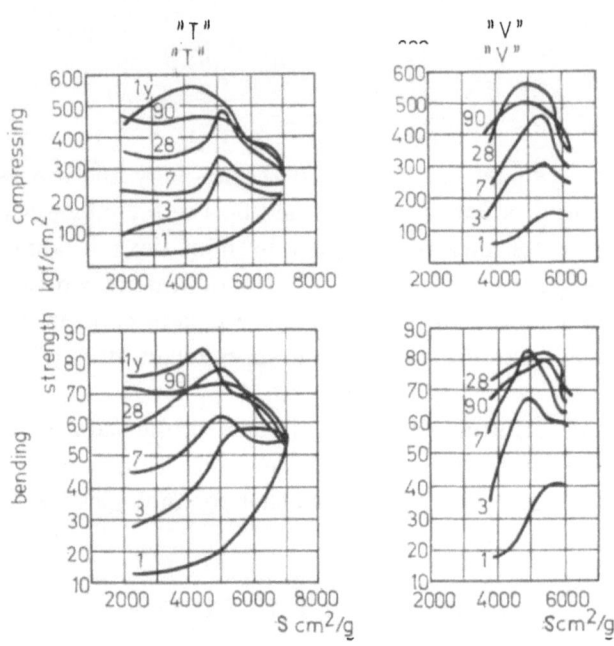

Fig. 19/1. Bending and compressing strength against Blaine surface for two cement samples

However, the final strength of cement is not determined, contrary to widely accepted opinions, by the absolute value of the specific surface—determined mostly by the Blaine method. Above a certain value of specific surface, approximately 5000 cm²/g Blaine, both the 28 days strength and the percentage of 3–30 microns quoted later as Δ (3–30) will decrease (Fig. 19/1). In Fig. 19/2, Δ (3–30) as a function of specific surface is shown, this latter is calculated according to formula (3.12). The similarity between Figs 19/1 and 19/2 is obvious.

Thus, we can formulate the object of cement grinding technology which is to achieve the maximum rate of Δ (3–30), whereas a sufficient rate of Δ (0–3) is always present due to the natural development of particle size distribution.

In Fig. 19/3 the percentages of Δ (3–30) and Δ (0–3) are plotted as functions of $R(0.09)$, with different n uniformity coefficients. In the case of 3–10% residue, the usual value with open circuit grinding, the best conditions can be obtained if $n = 1$ which is the customary value in commercial mills. If, however, a very high fineness is aimed at a higher uniformity coefficient (say $n = 1.2$) is convenient.

In Fig. 19/4 the percentage Δ (3–30) is plotted against Δ (0–3), which latter is determining for the specific surface. It is evident that the percentage Δ (3–30) increases with increasing n. More clearly expressed: *in the case of high uniformity coefficient a super duty cement can be produced having, however, a relatively low*

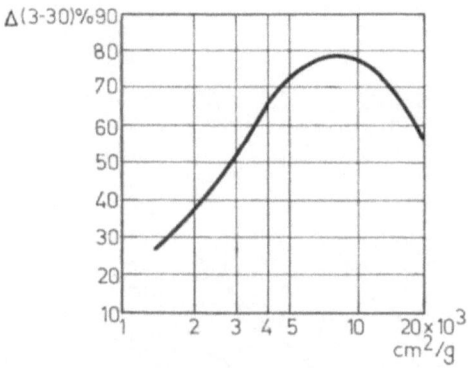

Fig. 19/2. Δ (3–30) against specific surface

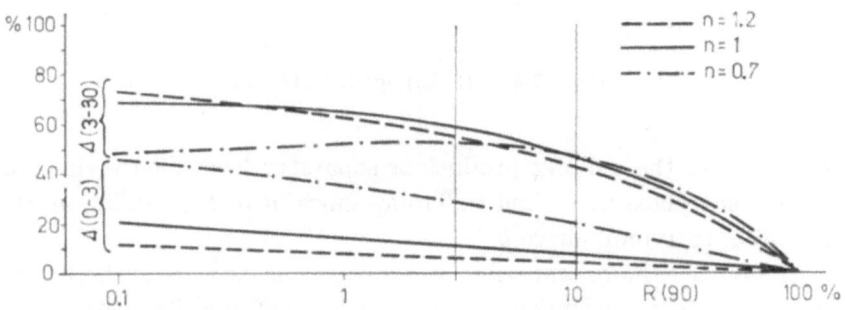

Fig. 19/3. Δ (3–30) and Δ (0–3) against R(0.09)

specific surface area (Beke 1973, Locher et al. 1973), in agreement with Rittinger's principle this means the possibility of energy saving. It is interesting to observe that in the case of a high uniformity coefficient, an arbitrary increase in specific surface is impossible due to lack of superfine particles.

Looking back to our former analysis of agglomeration (Chapter 4), it can be stated that a high value of uniformity coefficient in cement grinding yields three advantages: it retards agglomeration, it ensures favourable strengthening conditions, it allows a saving in energy consumption.

It has already been mentioned that there are two methods to obtain a high uniformity coefficient:

— very small grinding bodies in the finishing stage, as in the "minipebs" mill (Chapter 11)
— a closed circuit operation.

The second of these methods deserves detailed analysis. Figure 19/5 represents the U circuit coefficient, the quotient of throughput and output (see Chapter 10) as a function of $R(0.09)$ of the mill throughput or separator feed (quoted as d in Chapter 10), and further Δ (3–30) and Δ (0–3), all with different n uniform-

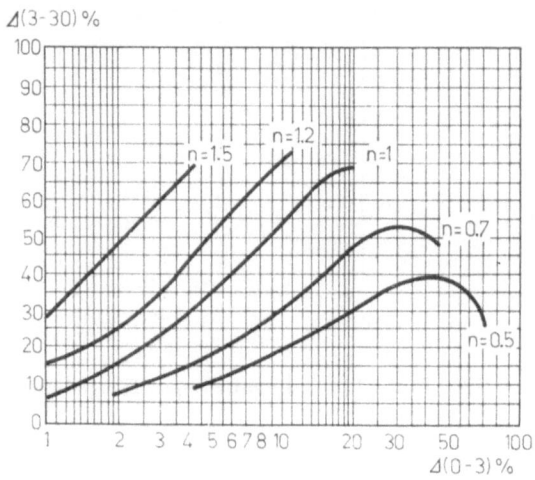

Fig. 19/4. Δ (3–30) against Δ (0–3)

ity coefficient of the milling product or separator feed. On the right side the separation takes place according to Tromp curve a in Fig. 10/3; on the left side, according to Tromp curve b.

Surprisingly, the more imperfect separation yields the far better results. The price is, of course, an increase in circulating load and its energy consumption—to be analysed later.

To explain this interesting result let us simulate the Tromp curves composed of straight sections forming a steplike diagram (Fig. 19/6) denoting by h the fictitious separating size, by v and w the respective separation rates in the fine and coarse sides. Figure 19/7 shows the effect of separation for a separator feed characterized by $n = 1$ and $R(0.09) = 30\%$; Δ (3–30) is, so to say, exclusively determined by w, the separating effect in the coarse side, v influences sooner the circuit number. Further analysis demonstrates that the separating size must be near $h = 30$ microns, even with case b in Fig. 10/3 or in Fig. 19/5. To enable us to obtain high w values, the increase of v is unavoidable.

The recycling of fine particles, i.e. increase of circulating load, decreases the capacity of the installation. A large classifier load at the same time causes the

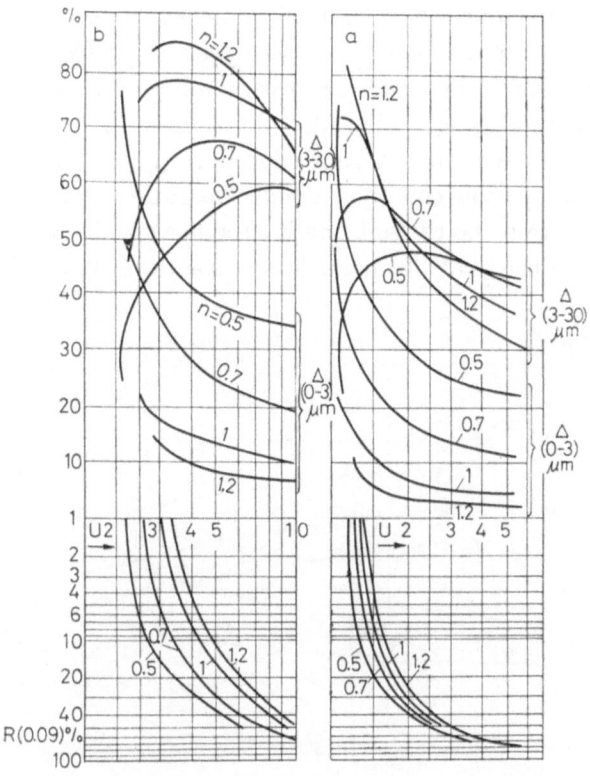

Fig. 19/5. Circuit coefficient, Δ (3–30) and Δ (0–3) as a function of $R(0.09)$

a and *b*: according to Tromp curves in Fig. 10/3

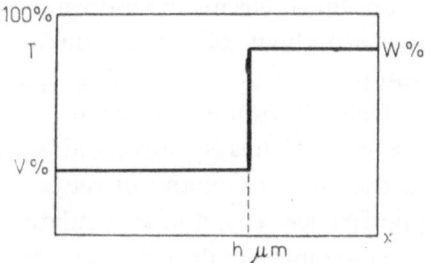

Fig. 19/6. Stepwise simulation of Tromp curve

selectivity of separation to deteriorate (see Chapter 18), thereby resulting in further increase of circuit number.

Here again the important role of the uniformity coefficient should be emphasized (Fig. 19/5): a lower n decreases the circulating load and the related energy consumption. The utilization of large grinding bodies results in a low uniformity coefficient which is advantageous from the viewpoint of the circulating load but disadvantageous from the viewpoint of particle size distribution.

To conclude our considerations we shall analyse the problem of energy consumption. Regarding the various and partly contradictory conditions, theorems of general validity cannot be established.

In the following, the conditions represented in Fig. 19/5 for $n = 1$ and $n = 0.7$ (and for comparison the open circuit operation) will be selected. To include in the comparison the cement quality too the specific energy consumption will be related to the Δ (3–30) particle size fraction. As our basis for com-

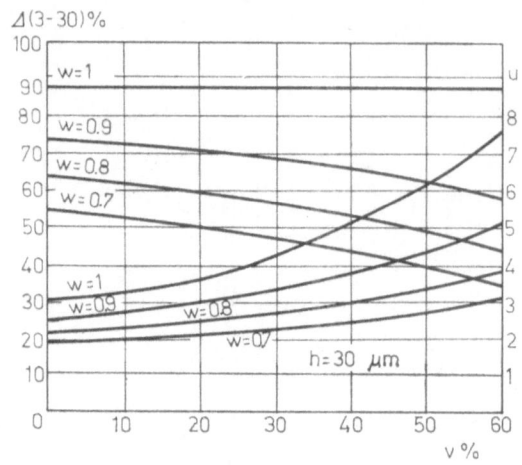

Fig. 19/7. Δ (3–30) and U as a function of v and w in Fig. 19/6. Separating size $h = 30$ microns

parison we shall take an open circuit operation characterized by $n = 1$, $R(0.09) = 8\%$, Blaine surface about 2800–3000 cm²/g, Δ (3–30)=49% having a practical energy consumption of about 35 kWh/t cement.

The energy demand of closed circuit operation is composed of two parts: the energy demand of the mill which is constant and independent of throughput or circulating load, and the energy demand of recycling. The output remains unchanged with changing fineness too, and is regulated by the function of the separator, resulting in a changing circulating load. The energy requirement of recycling is composed again of two parts: that of idling and a part increasing together with circuit coefficient. For simplicity and with tolerable approximation we shall calculate with 1 kWh/t throughput—which is equivalent to U kWh/t output. So the characteristic energy consumption of closed circuit cement grinding related to the fraction Δ (3–30) can be estimated by the formula

$$S = \frac{35 + U}{\Delta(3 - 30)} \quad \text{kWh/t.} \tag{19.1}$$

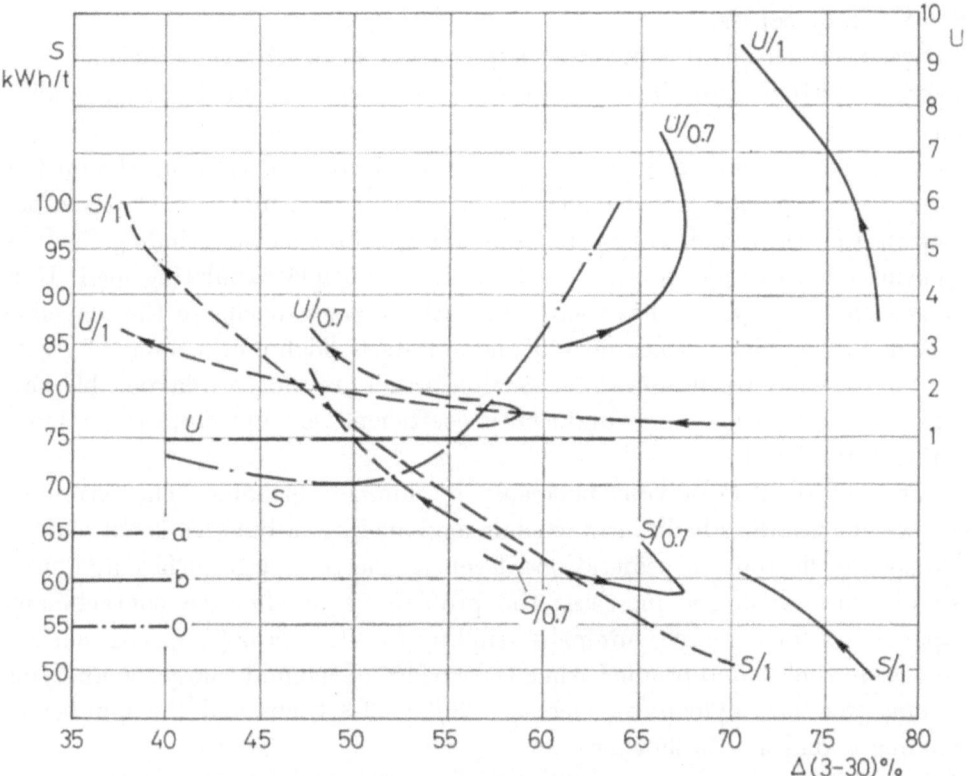

Fig. 19/8. Specific energy consumption related to fraction Δ (3–30) and circuit coefficient as a function of Δ (3–30)

a and *b*: according to Fig. 10/3; *o* open circuit

For open circuit operation the energy demand of the mill remains constant too but the output decreases with increasing fineness. This phenomenon can be taken into consideration using formula (4.8a). So we get

$$S = \frac{35}{\Delta(3 - 30)} \; \frac{q_R}{q_8} \quad \text{kWh/t.} \tag{19.2}$$

These relations are demonstrated in Fig. 19/8, specific energy consumption and circuit coefficient against Δ (3–30).

For super duty cements considerable energy saving by closed circuit is obvious; the numerical values are connected with the form of the Tromp curves. Particularly noteworthy is the loop with $n = 0.7$ hinting at an unstable process.

For open circuit, the line is extended till Δ (3–30) = 63% (equivalent $R(0.09) = 0.2\%$), a grinding fineness not to be achieved in conventional mill constructions. But the aforementioned "minipebs" mill (Cleemann 1972, see

Chapter 11) can produce super duty cements in open circuit by the application of 4–8 mm cylpebses.

It seems essential to look for an explanation as to why the efficiency of a closed circuit surpasses that of an open circuit when producing high quality cements.

In an open process, a higher fineness is attained by decreasing the output and as a consequence, increasing the specific energy consumption. On the other hand in a closed circuit the output remains constant as the grinding fineness is ensured by an appropriate classifying and multiple circulating load. It is indispensable, however, that there be a satisfactory capacity of the recycling system and the high permeability of the mill itself (high pore volume of grinding charge, wide diaphragm slots). Such a process results in a favourable particle size distribution (high uniformity coefficient, exempt from superfluous ultrafine particles).

But what about the favourable results by minipebs grinding? The extremely lightweight grinding bodies can exclude agglomeration, but this is not enough to disprove the above considerations. Even so, the very large surface and number of grinding bodies increases the probability of effective contacts and impacts. So, for example, minipebs grinding results in a 50% specific surface increase in the ground product while the surface of the mill charge — employing 20 mm or 6 mm cylpebses — increases $20/6 = 3.3$ times and the number of grinding bodies $3.3^3 = 36$ times.

For moderate quality cement, especially with small output (see Fig. 19/8), the more simple open process is preferred.

19.2 CEMENT RAW GRINDING

Concerning comprehensive data, let us again refer to the VDZ Congresses in 1971 and 1977 (Erni 1971, Sillem 1977) and Duda's Cement Data Book (1976–1977).

For the wet process, mostly three chamber compound mills in the open circuit are used. The charge of the first chamber consists sometimes of rods (rod-ped mill, Allis Chalmers, USA). Investigations with bowl- or rake classifiers and hydrocyclones remained unsuccessful because the moisture content of the slurry is limited to 35–40%. Wire screens are useless too, because of quick wire wear (sometimes in a few hours). Favourable raw material properties allow sometimes application of DSM arc screens (Fig. 18/5).

In the dry process drying-grinding installations are utilized as dealt with in Chapter 17.

19.3 LIME GRINDING

Hard burnt lime is ground in short, airswept ball mills; soft burned lime — as utilized in steel plants — is likely to aggregate and agglomerate therefore roller mills characterized by short retention time are required. Grinding aid dosing is advisable.

19.4 ORE GRINDING

Despite the immense industrial and economical importance of ore grinding (inclusive of bauxite grinding in alkalis) no special problems need to be solved. Grinding fineness is characterized by passing some one hundred microns; short ball mills serve to prepare for flotation or other wet concentration processes. Circuit rake classifiers or similar equipment, possibly hydrocyclones, are fitted. The trend indicates the use of very great capacities with driving powers reaching or surpassing 10 000 kW. Autogeneous grinding is more and more used.

Problems of ore grinding, including the trends of development, were set forth in detail in two papers by Tarján (1972, 1974).

19.5 COAL GRINDING

The required fineness is determined by the properties of the coal (calorific value, content of volatiles, moisture and ash), and by the firing system. The grinding method is to be chosen taking these into account. Ball mills, roller mills and impact mills are to be considered. This latter provides the coarsest product (20–40% residue on the 0.09 mm sieve), mostly applied in power works. For industrial kilns requiring higher temperatures, finer grinding is indispensable; roller mills or, rather, ball mills are convenient providing 12–15% oversize. Of course drying in the mill is essential. It should not be forgotten that measures to prevent explosions should be taken. Coals of high volatile content should not be permitted to dry below 5% moisture content, and an explosion valve needs to be built into the curvatures.

19.6 POTTERY AND FINE CERAMICS

The obsolete wet batch grinding is still used even now with flintstone or porcelain grinding media. Investigations hint at a particle size distribution for brittle components passing 40 microns and possibly no particles below 10 microns. It seems impossible to completely satisfy these requirements but in order to approach them a closed circuit with sharp separation is applicable. Special ceramics sometimes require extreme fineness to be satisfied by jet milling.

19.7 PHARMACEUTICALS

The object is to make it possible for the human organism to absorb insoluble substances. A very fine grinding is necessary with a feed of about 400–500 microns, a product maximum of about 20 microns. Small quantities are ground, the capacity being 20–50 kg/h. Mainly jet mills are applied.

19.8 MISCELLANEOUS

There are many other fine grinding operations of a very different nature not discussed above. As accentuated in the preface our considerations were restricted to brittle materials, for example, fertilizers to be ground in ball or roller mills; cosmetics, insecticides ground mostly in jet mills. These operations do not require theoretical considerations which were not dealt with in the previous chapters.

Symbols in Chapter 19

Δ (3–30)	percentage of size fraction 3–30 microns
Δ (0–3)	percentage of size fraction 0–3 microns
U	circuit coefficient
R	sieve residue
n	uniformity coefficient
h	separating size, microns
v, w	separating rate in the fine and coarse sides, respectively
S	specific energy demand related to Δ (3–30), kWh/t
q	output factor

CHAPTER 20

PERSPECTIVES

Our considerations should not be concluded, however, even though this is somewhat hazardous in view of the rapid advance in technology, without looking at the trends of development.

Every development must satisfy the following requirements:

- increasing grinding capacity
- saving in investment costs
- energy saving
- high availability
- complying with increasing quality demands
- introduction of computer control—not be dealt with in this monograph.

Analysis of the first mentioned two requirements seems to be inseparable. To enable us to obtain coherent data for comparison, capacities will be characterized by the kW driving power.

The biggest capacities attained nowadays in ore grinding with autogeneous mills exceeds 15 000 kW; in cement grinding with ball mills the figure is about 7000 kW; in cement raw grinding with ball mills or roller mills it is also about 7000 kW; in coal grinding with shockwheel or ball mills somewhat more than 1000 kW.

Investment costs can be accepted as approximately proportional to machine masses.

Based on catalogue data of manufacturers in Figs 20/1, 20/2 and 20/3, mass (weight) and specific mass values against driving power are presented for ball, roller and shockwheel mills (extrapolations are shown with dashed lines). As for autogeneous mills, their specific mass is some 50–60% bigger than those of conventional ball mills but of course no grinding body charge must be added.

Progress in mechanical engineering allows an increase in these capacities, viz. special transport trucks, welding on site, the driving of ball mills and roller

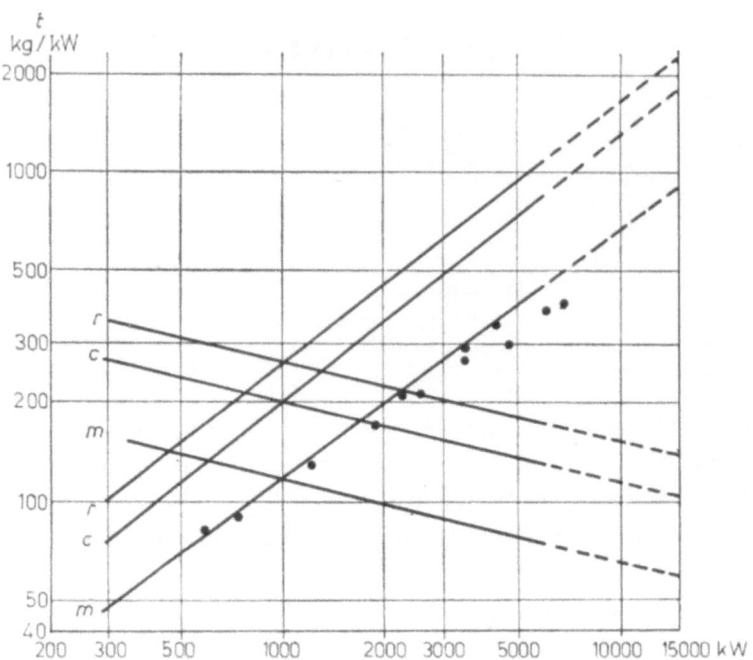

Fig. 20/1. Weight t, and specific weight t/kW of ball mill; m mill (excluded gear); c charge; r recirculation

mills by low frequency ring motors, but above all, by using wear-resistant construction materials.

Conspicuous for ball mills is the decreasing specific mass with increasing driving power, whereas with roller and impact mills it remains virtually unchanged.

On the other hand, the outer dimensions of ball mills considerably exceed those of roller and impact mills with the attendant high building costs. And, for example, the ball charge of a 15 000 kW ball mill approaches the frightening mass of 1000 t. The handling of these immense masses (change of ball charge) will inevitably have a detrimental effect on the availability factor.

And — as mentioned in Chapter 12 (p. 81) — the influence of the voids between the grinding bodies cannot be neglected since it has a limiting effect on the increase in ball mill sizes. This limit is, however, beyond those sizes of importance in present times or, for that matter, in the foreseeable future.

For cement grinding an upper limit is given by the rise of temperature, as mentioned in Chapter 12. Investigating this problem in detail Reuss (1974) indicated 7 m as the maximum value for the diameter. This corresponds to about 20 000 kW or 550 t/h, i.e. far above present-day requirements. For ore grinding in autogeneous mills no such upper limit exists.

134

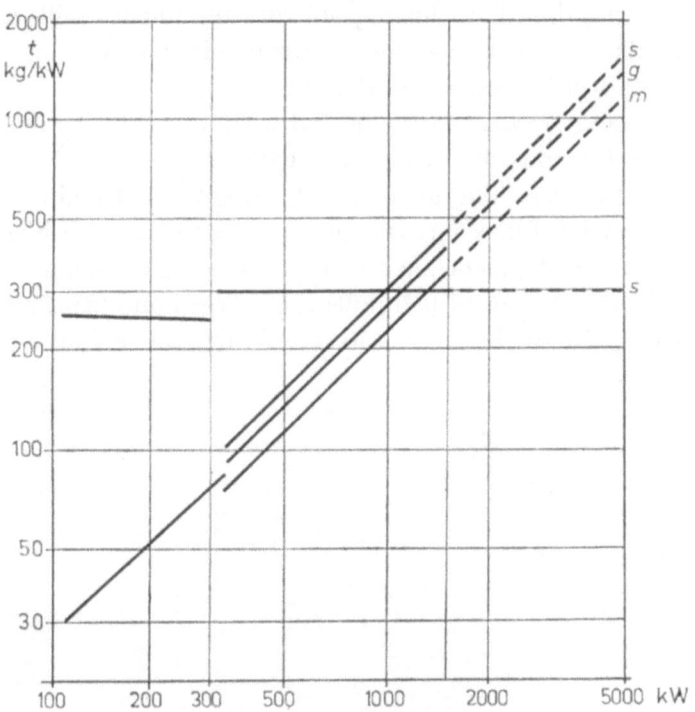

Fig. 20/2. Weight and specific weight of roller mills
m mill; *g* gear; *s* separator

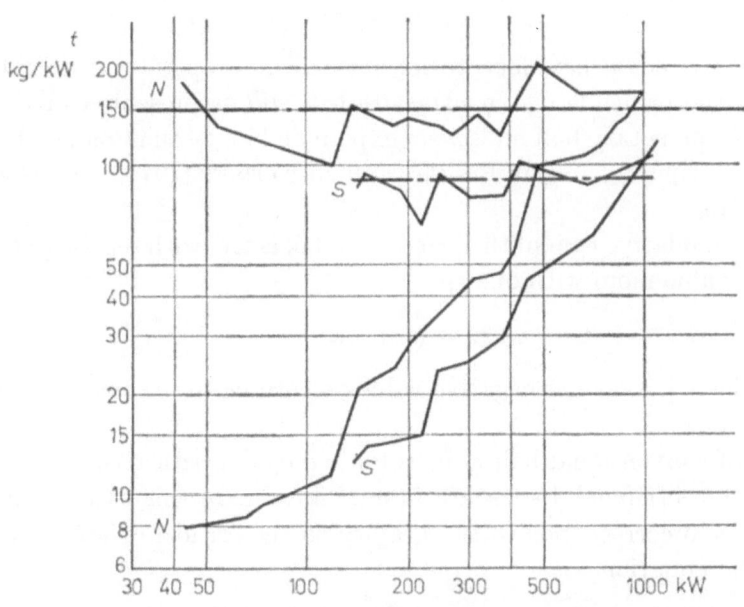

Fig. 20/3. Weight and specific weight of shock wheel mills
S light, *N* heavy series

For very fine grinding both high speed rotating impact mills and jet mills are in steady development, and quality demands, and the operational and investment costs will be decisive in the choice.

Problems of energy saving are connected with technological requirements — dealt with in the previous chapter — and helped by the application of computer control in regard to searching for and stabilizing the optimal conditions.

To conclude our brief look into the future, three mill types which are now in an initial state of development must be described and considered.

To reduce the dimensions of ball mills the energy concentration must be increased. With regard to the small and unalterable value of the gravitational

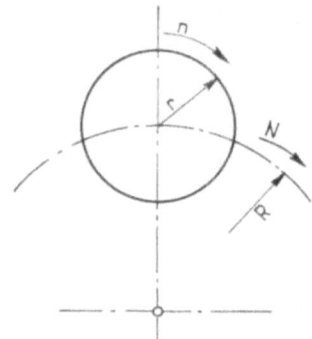

Fig. 20/4. Principle of planetary ball mill

r is the radius; n the rpm of the grinding tube; R and N those of gyration; z the sun-to-planet gear ratio

acceleration it is not a new idea to let the ball mill operate in a centrifugal field. This is the principle of the *planetary ball mill* as presented in Fig. 20/4.

Processes in planetary ball mills were expounded in detail in papers by Joisel (1956) for dry grinding, and by Bradley and co-workers (1971, and 1974–1975) for wet grinding.

The own, i.e. relative rpm of the grinding tube, is zN, with regard to gyration the absolute value more with one rpm

$$n = (z + 1)N \tag{20.1}$$

where z can have positive or negative values according to the course of grinding tube rotation.

Capacity of conventional ball mills is limited by the critical rpm where gravitational and centrifugal forces are in equilibrium. In this case the gyration centrifugal force corresponds to the gravity so the critical condition is determined by the equation

$$R \left(\frac{N\pi}{30} \right)^2 = r \left(\frac{n\pi}{30} \right)^2 = r \left[\frac{(z + 1) N\pi}{30} \right]^2. \tag{20.2}$$

136

Somewhat strikingly N falls out, critical condition is independent of rotation or gyration speed and is determined only by their ratio, or the ratio of their radii:

$$z_{kr} = -1 \pm \sqrt{\frac{R}{r}}. \tag{20.3}$$

It is interesting to observe the case $z = -1$ where no critical state takes place—excluding R=0, the conventional ball mill—and the absolute rpm is equal zero (Fig. 20/5).

Energy demand or output accepted as proportional to energy demand can be calculated according to Bradley

$$P \sim pr^3 \, Lz \, RN^3 \tag{20.4}$$

and according to Joisel

$$P \sim pr^4 \, LN^3 \left(1.54 + 0.2 \, \frac{R}{r}\right) \tag{20.5}$$

where P output t/h, p the number of grinding tubes, L the tube length, R, r and N as above.

Equations (20.3), (20.4) and (20.5) are of immense importance: capacity of planetary ball mills is proportional to the cube of gyration speed and this speed is not limited by a critical upper value. Consequently with small dimensions surprisingly high output can be attained. According to calculations by Joisel a cement mill of the output of 30 t/h can be characterized by $p = 6$, $L = 100$ cm, $R = 30$ cm, $N = 320$/min, the space demand about 1 m³, the

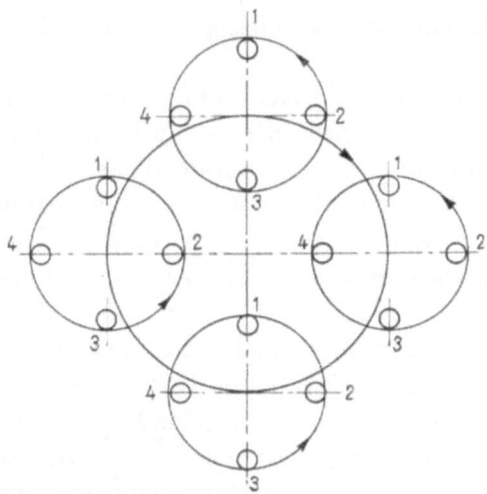

Fig. 20/5. Performance of planetary ball mill for $z = -1$

centrifugal acceleration 337 m/s², about 34 g. For comparison: a conventional ball mill of this capacity has the main dimensions 3 dia times 10 m.

But of course the speed-increase has its limits too: the mechanical stress in spokes, bearings, gears, etc.

Bradley and co-workers published test results and predicted performance of a gold ore wet grinding three tube prototype planetary mill with the main data: $r = 10$ cm, $R = 16.5$ cm, $L = 35$ cm, $z = -1$, and variable speed between 4.2 and 16.6 Hz. Centrifugal acceleration extended as far as 100 g, net power 175 kW, the mill charge was about 75 kg. A conventional ball mill of 175 kW has a charge of about 20 t. Grinding body size seemed to be proportional to $1/\sqrt{N}$.

All these very hopeful performances are consequences of the high energy concentration which on the other hand results in quick wear and high heat generation.

According to Joisel's estimation the wear surpasses by two orders of magnitude those with conventional mills, whereas with quartzite wet grinding Bradley stated 4 kg/t grinding body wear and 1 kg/t liner wear. Better results were attained with rubber liner and a quite new idea is the application of pneumatic liners.

Here again we must emphasize that: the small masses allow quick change of charge or even the whole tube, this latter required some three hours.

As for heat generation in dry grinding no reliable experiences are available. But in consequence of high gyration speed good heat transfer is assured, the specific tube surface is high and cooling ribs or even water sprinkling can be applied.

Although the planetary ball mills are now in early stage of development, their important role in the future seems as certain.

Another quite different kind of milling operations is the *shock shatter* or *Snyder process*.

Tensing strength of brittle materials is only a few percentage of their compressive strength. So it is not a new idea to try to attain breakage by tension which, of course, is impossible by conventional mechanical methods. But in the thirties there were known experiments to set the solid body into a pressure vessel where the high pressure steam or gas permeates the pores of the material and then on sudden release of pressure flashes explosively, thus shattering the material.

Recently Synder succeeded in perfecting the operation to commercial scale for application in ore grinding (Cavanaugh 1972, 1973).

A scheme of the Snyder process is presented in Fig. 20/6. From bin 1 the material is fed periodically into pressure vessel 3. The process begins with the fluidization of this material bed with a working fluid, in Snyder's investigations 6–30 atm saturated steam. Flow is initiated by opening a special quick-opening valve (6) — hitherto unknown machine element — into the discharge duct. As this valve is opened, *in less than 15 ms*, the compressible working fluid expands and the previously fluidized particles are accelerated and entrained with supersonic or near sonic speed through the convergent-divergent duct into the collecting chamber. Here sometimes an impact plate (8) is placed.

As evident: the milling process is composed of two parts: tensile stress by abrupt expansion of the fluid and the subsequent short period jet or impact grinding.

To attain continuous operation the pressure vessel must be dosed with material in regular intervals, according to references in 15–30 seconds. But the

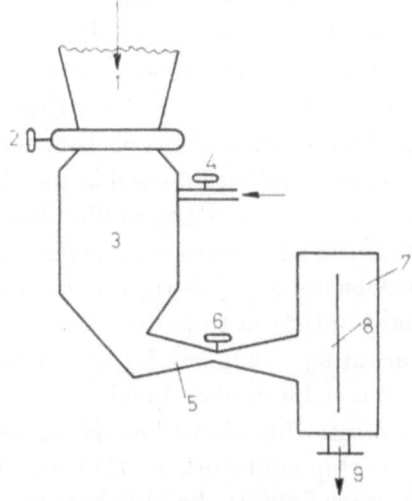

Fig. 20/6. Principle of Snyder mill

1 bin; 2 loading valve; 3 pressure vessel; 4 pressure valve; 5 convergent-divergent duct; 6 quick valve; 7 collecting vessel; 8 impact plate; 9 material outlet

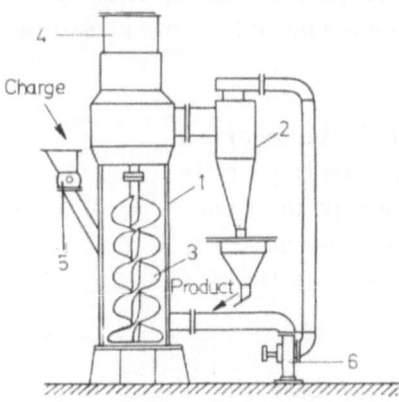

Fig. 20/7. Dry type tower mill (Japan Tower Mill Co. Ltd., Tokyo)

1 main body; 2 cyclone; 3 screw; 4 geared motor; 5 feeding rotor; 6 blower

crucial element is the above mentioned fast valve operator of the operating time of 15 ms.

The first period produces particles separated on the natural grain boundaries, essential in ore dressing and the subsequent jet milling will accomplish no decisive further size reduction: so there are mostly particles of the size 0.2–0.3 mm produced.

As it is quite clear, the Snyder process is a selective grinding process for moderate fineness requirements, fitted into the flowsheet of flotation ore dressing.

In Fig. 20/7 a dry *tower mill* is illustrated. This is a new mill type imitating the principle of mutual abrasion of rocks in a river bed. Wet and dry types have been developed. This mill has an upright cylindrical main body, a cyclone to gather the particles and a blower; these are all linked with ducts to form a closed circuit. The main body is charged with grinding media such as steel balls, hard pebbles, ceramic balls, etc. The screw is driven by a vertical geared motor. The feed is ground by mutual abrasion between material charge and grinding media. Autogeneous grinding is possible too. In the wet process the cyclone and blower are replaced by a hydraulic classifier and sand pump. According to the manufacturer's leaflet driving capacities from 0.75 to 150 kW are available; the dimensions (height, length, width) are from $3.3 \times 2.2 \times 1.2$ to $8.2 \times 5 \times 3$ m. The standard output is 10 to 16 000 kg/h in dry grinding -150 mesh (0.1 mm); in wet grinding -325 mesh (0.045 mm). To avoid wear, it is recommended that the screw be rubber lined.

To obtain high capacities, the above described new grinding methods require thorough-going development work so they can be established: in the following 10–20 years the nowadays applied kinds of machineries and processes will maintain their leading role.

To conclude this chapter and the whole monograph: the most important requirement of all further progress in the field of grinding operations is the use or moreover the development of high wear resistant construction materials.

Symbols in Chapter 20

r radius of grinding tube m or cm
n speed of grinding tube rpm or Hz
L length of grinding tube m or cm
R radius of gyration m or cm
N speed of gyration rpm or Hz
z gear ratio
P output t/h
p number of grinding tubes

REFERENCES

Akunow, W. I. and Deschko, Ju. I. (1978): Entwicklungstendenzen der Zementrohmaterial- und Klinkermahlung. *Zement-Kalk-Gips* **31** 10–11

(Alyavdin, V. V.) Алявдин, В. В. (1938): Процесс размола в трубных мельницах, Сборник „Вопросы помола в цементной промышленности". Изд-во Гипроцемента. *(Process of Grinding in Tube Mills*, in the collection "Problems of Grinding in the Cement Industry)"

Anselm, W. (1950): *Zerkleinerungstechnik und Staub*, VDI Verlag, Düsseldorf

Austin, L. G. and Klimpel, R. R. (1964): Theory of Grinding Operations. *Ind. Eng. Chem*, **56** 18–29

Bachmann, D. (1940): *Die Bewegungsvorgänge in Schwingmühlen mit trockener Mahlkörperfüllung*. Zeitschr. VDI Beiheft Verfahrenstechnik 43–55

Bernhardt, C. and Heegn, H. (1975): *Zur mechanischen Aktivierung in Feinzerkleinerungsmaschinen*. Fourth European Symp. Comminution, Preprint 213–225

Bernutat, P. (1961): Die Berechnung der Mahlbarkeit im Mahlbarkeitsprüfer. *Tonindustrie-Zeitung* **85** 397–398

Bernutat, P. (1964): Verschleiss von Mahlkörpern und Mahlplatten. *Zement-Kalk-Gips* **17** 397–400

Bombled, J. P. (1967): Recherches des dimensions optimales et de l'échelonnement des corps broyants dans les broyeurs à boulets. 2. Symposium Européen sur la fragmentation, *Dechema Monographien* **57** 633–665

Bombled, J. P. (1972): L'usure des corps broyants dans les broyeurs à boulets. 3. Symposium Européen sur la fragmentation. *Dechema Monographien* **69** 843–880

Bond, F. C. (1952): The Third Theory of Comminution, *Mining Engineering* 484–494 (AIME Trans. **193**)

Bond, F. C. (1958): Grinding Ball Size Selection. *Mining Engineering* 592 ff. (AIME Trans. **211**)

Bond, F. C. (1961): Crushing and Grinding Calculations. *British Chemical Engineering* **6** (June)

Bond, F. C. and Wang, J. T. (1950): A New Theory of Comminution. *Mining Engineering* 871–878 (AIME Trans. **187**)

Boross, L. (1969): Leistungsbedarf von Ventilatormühlen. *Acta Techn. Ac. Sci. Hung.* **65** 67–78

Börner, H. (1965): Mahlkörperzusammensetzung in Rohrmühlen der Zement- und Kalkindustrie. *Zement-Kalk-Gips* **18** 420–428

Bradley, A. A. (1971): *Some Principles of Centrifugal Milling*. 3rd. Symp. on Comminution, Cannes. Preprint 705–724

Bradley, A. A., Feemantle, A. J. and Lloyd, P. J. D. (1974): *J. South African Inst. Mining and Metallurgy* 379–387 loc. cit. (1975): 78–80

Broadbent, S. R. and Callcott, T. G. (1956): Coal Breakage Processes. *J. Inst. of Fuel* **29** 524–539

Brown, M. A. (1959): Calculation of Closed-Circuit Grinding. *British Chemical Engineering* 463–466

Callcott, T. G. (1968): Size Distribution and Single Particle Breakage. The *B. H. P. Technical Bulletin* **26** 26–30

Cavanaugh, W. J. (1972): Snyder Process — a Breakthrough in Comminution. *Mining Congress Journal*, Dec. 30–36

Cavanaugh, W. J. and Rogers, D. J. (1973): *Applications of the Snyder Process*. Tenth Int. Mineral Processing Congress, London. Paper 46

Charles, R. T. (1957): Energy–Size Reduction Relationships in Comminution. *Mining Engineering* 80–88 (AIME Trans. **208**)

Chujo, K. (1949): Study on Comminution. *Ann. Report Soc. Japan Chem. Engrs.* **7** 1–83

Cleemann, J. (1972): Entwicklung der Durchlaufmahlung. *Zement-Kalk-Gips* **25** 63–66

Davies, R. (1973–1974): Rapid Response Instrumentation for Particle Size Analyses. *American Laboratory*, Dec. 17–23; Jan. 73–86; Feb. 47–55

Davis, E. W. (1920): Fine Crushing in Ball Mills. *Transactions AIMME* Vol. LXI. 250–296

Deckers, M. (1972): Über die Mahlbarkeit von Zementklinker. *Zement-Kalk-Gips* **25** 445–448

(Deshko, Yu. I., Akunov, V. J., Pankratov, V. L., Ferens, N. I. and Kolosovskaya, V. N.), Дешко, Ю. И., Акунов, В. И., Панкратов, В. Л., Ференс, Н. И., Колосовская, В. М. (1973): Струйный помол повышает качество цемента (Jet milling improves Cement Quality), Цемент **5** 16–17.

Drohsihn, U. (1972): Verschleiss in Rohrmühlen. *Zement-Kalk-Gips* **25** 571–574

Duda, F. W. (1976 or 1977): Cement Data Book. Bauverlag GmbH Wiesbaden and Berlin

Eifel, M. and Schönert, K. (1976): Untersuchungen zur Mahlung in Kugelmühlen mit glatten und rillenprofilierten Panzerplatten. *Zement-Kalk-Gips* **29** 30–36

Ellerbrock, H. G. (1975): *Über die Mahlbarkeitsprüfung von Zementklinker*. Fourth European Symp. Comminution. Preprint 197–211

Erni, H. (1971): Rohmaterialaufbereitung und Homogenisierung. *Zement-Kalk-Gips* **24** 487–496

Fasbender, H. (1971): Die Tandem-Mahlanlage — ein modernes Mahltrocknungssystem für Zementrohmaterial. *Zement-Kalk-Gips* **24** 499–502

Fáy, G. and Zselev, B. (1960): Die mathematischen Grundlagen der Körnungsanalyse von gemahlenen Kornmengen. *Acta Techn. Acad. Sc. Hung.* **44** 3–4, 237–260

Furuya, M., Nakajima, Y. and Tanaka, T. (1971): *Ind. Eng. Chem. Process Des. Develop.* **10** 449–456

Gebelein, H. (1956): Beiträge zum Problem der Kornverteilungen. *Chemie Ingenieur Technik* **28** 773–782

Gildemeister, H. H. and Schönert, K. (1975): *Bruchphänomene und Spannungsfeld in prallbeanspruchten Kugeln*. Fourth European Symp. Comminution. Preprint 131–149

Götte, A. (1952): Fragen der Hartzerkleinerung. *Zement-Kalk-Gips* **5** 383–394

Griffith, A. A. (1921): The Phenomena of Rupture and Flow of Solids. *Phil. Trans. Roy. Soc. A.* **221** 163–198

Haese, U., Scheffler, P. and Fasbender, H. (1975): Mahlbarkeitsprüfung und Rohrmühlenauslegung bei Zementrohstoffen. *Zement-Kalk-Gips* **28** 316–324

Halbart, G. and Freymann, V. (1955–1956): Le revêtement du broyeurs à boulets. *Revue des Matériaux de Construction* **481** loc. cit. **484**

Harris, C. C. (1967): On the Limit of Comminution. *Soc. Mining Engineering*, March, 17–30

Heidenreich, H. (1954): *Die Erfolgsrechnung im Aufbereitungsbetrieb*. Verlag Glückauf GmbH, Essen

Hint, J. (1976): *Über die grundlegenden Probleme der mechanischen Aktivierung*. Institut für wissenschaftlich-technische Information und ökonomische Forschungen der Estnischen SSR

Hiorns, F. J. (1967): Energy Conversion in Milling. 2nd Symp. on Comminution, *Dechema Monographien* **57** 165–181

Holmes, J. A. (1957): A Contribution to the Study of Comminution. *Trans. Inst. Chem. Engrs.* **35** 125–156

Hukki, R. T. (1958): *Grinding at Supercritical Speeds in Rod and Ball Mills*. Trans. Int. Mineral Dressing Congress, Stockholm, 85–122

Hukki, R. T. and Reddy, I. G. (1967): The Relationship between net Energy Input and Fineness in Comminution. 2nd European Symp. Comminution. *Dechema Monographien* **57** 313–339

Hüttig, G. F. (1952): Neue Beobachtungen bei Zermahlungsvorgängen und deren Deutung. *Dechema Monographien* **245–248** 96–115

Iwabuchi, T. (1968): *A Concept of the Cause of Agglomeration and Mechanism of Grinding Aid Action in Cement Grinding*. The Cement Association of Japan, 22nd General Meeting. 95–99

Iwata, H., Masuda, K. and Yamamoto Kawaguchi, M. (1974): Über die Kugelfüllung bei Schwingzerkleinerung in lotrecht schwingendem Gefäss. *Aufbereitungs-Technik* 553–557

Jäger, H. (1976): Zementmahlverfahren mit mehreren Sichtern. *Zement-Kalk-Gips* **29** 6–11

Jimbo, G. (1972): Recent Work in Japan on Size Reduction. *J. Res. Assoc. Powder Tech.* **9** 73–80

Jimbo, G. and Suzuki, J. (1973): The Effect of Gas Atmosphere on the Grinding Efficiency of Limestone in Vibration Ball Mill. *J. Soc. Mat. Sci. Japan*, Vol. 22, **138** 697–701

Joisel, A. (1956): Broyeurs à satellites. *Revue des Matériaux de Construction* **493** 234–250

Joisel, A. and Birebent, A. (1951–1952): Mécanique interne du broyeur à boulets. *Revue des Matériaux de Construction* 434–439

Juhász, Z. (1972) Mechanokémiai reakciók (Mechanochemical Reactions). (In Hungarian.) *Építőanyag* **24** 365–367

Juhász, Z. (1974): Mechanochemische Aktivierung von Silikatmineralien durch Trocken-Feinmahlen. *Aufbereitungs-Technik* 558–562

Kannewurf, A. S. (1957): Research pushes Grindability Guesses into the Background. Rock Products May 66 ff. Ref: *Zement-Kalk-Gips* **1957** 435

Kemmann, W. (1972): Richtlinien für die Zusammenstellung einer Mahlkörperfüllung für eine Rohrmühle. *Aufbereitungs-Technik* 313–318

Kihlstedt, P. G. (1962): *The Relationship between Particle Size Distribution and Specific Surface in Comminution*. Symp. Zerkleinern Verlag Chemie, Weinheim, VDI Verlag, Düsseldorf, 205–216

Klovers, E. J. (1973): *Performance of Large Roller Mills*. Rock Products 9th International Cement Industry Seminar, Chicago

Kolostori, J. (1975): Verschleissfunktion der Kugelmühlengattierung und ihre Feststellung in der Praxis. *Silikattechnik* **26** 205–209

Korda, P. (1961): Strahlzerkleinerung und Trocknung. *Aufbereitungs-Technik* 230–239

143

(Kovaljukh, V. R. and Gud, M. B.) Ковалюх, В. Р., Гуд, М. Б. (1978): Зависмость производительности цементноймельницы от коэффициента заполнения и частоты вращения. (Output change of cement mills as a function of filling rate and rotation speed) Цемент **7**. 8–9.

Lehmann, H. and Haese, U. (1955): Der Mahlbarkeitsprüfer, ein Gerät zur Untersuchung der Mahleigenschaften harter Stoffe. *Tonindustriezeitung* **79** 91–94

Leschonski, K., Alex, W. and Koglin, B. (1974–1975): Teilchengrössenanalyse. *Chemie-Ing.-Techn.* **46** 23–26, 101–106, 289–292, 387–390, 477–480, 563–566, 641–644, 729–732, 821–824, 901–904, 984–987 **47** 97–100

Locher, F. W., Sprung, S. and Korf, P. (1973): Der Einfluss der Korngrössenverteilung auf die Festigkeit von Portlandzement. *Zement-Kalk-Gips* **26** 349–355

Menyhárt, J. (1972): Mechanokémiai folyamatok (Mechanochemical Processes). (In Hungarian.) *Magyar Kémikusok Lapja* **31** 520–526

Menyhárt, J., Domsa, K. and Németh, J. (1955): Hungarian patent No. 153843. Ref.: *Dechema Monographien* **57** 491

Mori, Y. (1962): Studies on Jet Pulverizing. Symp. Zerkleinern, Verlag Chemie, Weinheim, VDI Verlag, Düsseldorf, 515–530

Németh, J. and Horányi, R. (1970): Untersuchungen über die Teilchengrösse als Kennwert der Leistung schneller Klärzentrifugen. *Periodica Polytechnica Ch.* XIV/2, Budapest, 183–193

Nudel, M. E. and Krichtin, G. Sz. (1976): Felületaktív anyagok hatása cementipari nyersanyagok őrlésére (Effect of Surface Active Agents on the Grinding of Cement Raw Materials). (In Hungarian.) *Építőanyag* **28** 397–405

Ocepek, D. and Eberl, E. (1975): *Agglomeration oder Rehbindereffekt.* Fourth European Symp. Comminution. Preprint 183–195

Okuda, S. (1971): Das Zerkleinerungsverhalten spröder Stoffe bei der Prallzerkleinerung. *Aufbereitungs-Technik* 529–536

Olivero, L., Caire, F., Pintor, G. and Bianchi, P. (1977): Betriebliche Anpassung von Zementmühlen mit variabler Drehzahl. *Zement-Kalk-Gips* **30** 644–646

Opoczky, L. and Mrakovics, K. (1976): Über die Rolle des Gefüges und der chemisch-mineralogischen Zusammensetzung des Zementklinkers bei der Grob- und Feinmahlung. *Freiberger Forschungshefte* **A 553** 71–81

Opoczky, L. (1977): Fine Grinding and Agglomeration of Silicates. *Powder Technology* Vol. **17** No. 1. 1–7

Opoczky, L. (1978): Personal communication.

Papadakis, M. (1960): Contribution à l'étude des broyeurs à boulets industriels. *Revue des Matériaux de Construction* **542** 295–308

Paulsen, H. (1964): Mahlkörperverteilung in Rohrmühlen. *Zement-Kalk-Gips* 392–396

Perow, V. A. and Brand, J. W. (1954): *Feinmahlen der Erze.* VEB Verlag Technik, Berlin. (Originally published in Russian. Moscow, 1950.)

Peters, K. (1962): *Mechanochemische Reaktionen.* Symp. Zerkleinern. Verlag Chemie, Weinheim, VDI Verlag, Düsseldorf, 78–98

Pethő, S. (1971): Neue Kennwerte zur Beurteilung von Trennvorgängen. *Aufbereitungs-Technik* 743–745

Pethő, S. and Tompos, E. (1974): About the New Index Numbers of Separation. *Acta Techn. Ac. Sci. Hung.* **78** 237–256

Pethő, S. and Smirnow, S. (1976): Übergangsfunktionen und Kennwerte aufbereitungstechnischer Trennvorgänge. *Glückauf-Forschungshefte* **37** 216–224

Planiol, R. (1962): *Les broyeurs centrifuges et le vide.* Symp. Zerkleinern, Verlag Chemie, Weinheim, VDI Verlag, Düsseldorf, 503–514

144

Rehbinder, P. (1944): *Hardness Reducer in Drilling*. (Transl. from Russian). Publ. by Council for Scientific and Industrial Research, Melbourne, 1948

Rehbinder, P. A. and Chodakow, G. S. (1962): Feinmahlung von Quarz. *Silikattechnik* **13** 200–208

Reményi, K. (1966): Investigation on Grindability of Limestone and Rock-Salt Mixture in Hardgrove Mill. *Acta Techn. Ac. Sci. Hung.* **56** 75–90

Reuss, A. K. (1974): Die thermische Belastungsgrenze bei der Zementvermahlung. *Zement-Kalk-Gips* **27** 321–329

Richter, L., Bornschein, G. and Scheibe, W. (1974): Überleitungsergebnisse beim Einsatz von Mahlhilfsmitteln in der Zementindustrie. *Silikattechnik* **25** 399–401

Rose, H. E. (1962): *Some Observations on Vibration Mills and Vibration Milling*. Symp. Zerkleinern, Verlag Chemie, Weinheim, VDI Verlag, Düsseldorf, 427–454

Rose, H. E. (1967): A Comprehensive Theory of the Comminution Process. 2. Symp. on Comminution. *Dechema Monographien* **57** 27–62

Rowland, C. A. (1972): *Large Grinding Mills*. Intermountain Minerals Conference, Vail, Colorado

Rowland, C. A. (1975): *The Tools of Power Power*. AIME Annual Meeting, Arizona Section

Rumpf, H. (1959): Beanspruchungstheorie der Prallzerkleinerung. *Chemie Ing. Techn.* **31** 323–337

Rumpf, H. (1960): Prinzipien der Prallzerkleinerung und ihre Anwendung bei der Strahlmahlung. *Chemie Ing. Techn.* **32** 129–135

Rumpf, H. (1962): *Über grundlegende physikalische Probleme bei der Zerkleinerung*. Symp. Zerkleinern. Verlag Chemie, Weinheim, VDI Verlag, Düsseldorf, 1–30

Rumpf, H. (1965): Über die Feinheitsbestimmung von technischen Stäuben. *Staub* **25** 15–22

Rumpf, H., Faulhauber, F., Schönert, K. and Umhauer, H. (1967): Analyse der Brucherscheinungen in Glaskugeln und kreisrunden Glasscheiben. 2. Symp. on Comminution. *Dechema Monographien* **57** 85–126

Schardin, H. (1962): Kinematographische Analyse des Bruchvorganges. Symp. Zerkleinern, Verlag Chemie, Weinheim, VDI Verlag, Düsseldorf, 31–48

Schauer, S. (1977): Zementmahlung mit MPS-Walzenschüsselmühlen. *Zement-Kalk-Gips* **30** 576–578

Scheibe, W., Dombrowe, H. and Herrmann, R. (1975): Die Beeinflussung der Haftkräfte in Mahlprodukten durch Mahlhilfsmittel. *Silikattechnik* **26** 243–245

Schellinger, W. (1952): Solid Surface Energy and Calorimetric Determinations of Surface Energy Relationships for Some Common Minerals. *Mining Engineering* 369–374

Schildknecht, W. (1978): Betriebserfahrungen mit grossen Wälzmühlen. *Zement-Kalk-Gips* **31** 1–2

Schönert, K. and Steier, K. (1971): Die Grenze der Zerkleinerung bei kleinen Korngrössen. *Chemie-Ing. Techn.* **43** 773–777

Schrader, R. and Dusdorf, W. (1966): Die mechanische Aktivierung von Quarz. *Kristall und Technik* **59** ff

Schrader, R., Hoffmann, B., Plänitz, H. and Heuneberger, J. (1970): Über aktiviertes Calciumoxid. *Zement-Kalk-Gips* **23** 194–199

Schramm, R. and Gaitsch, E. (1974): Eine quantitative Methode zur Gattierung in Kugelmühlen. *Zement-Kalk-Gips* **27** 330–332

Seebach, H. M. (1969): *Die Wirkung von Dämpfen organischer Flüssigkeiten bei der Zerkleinerung von Zementklinker in Trommelmühlen*. Schriftenreine der Zementindustrie, 35 VDZ Düsseldorf. See also *Zement-Kalk-Gips* **22** 202–211

Sillem, H. (1972): *Mahlen und Lagern von Klinker und Zement.* VDZ Kongress '71. Bauverlag GmbH Wiesbaden, 187–196. See also *Zement-Kalk-Gips* **25** 53–62

Sillem, H. (1977): Zerkleinerungstechnik, *Zement-Kalk-Gips* **30** 549–557

Slegten, J. (1964): Betrachtungen über Mahlkörper und Panzerungen. *Zement-Kalk-Gips* **17** 503–510

Smekal, A. (1937): Grundvorgänge der Hartzerkleinerung. *Zeitschr. VDI* **81** 1321–1326

Stairmand, C. J. (1975): *The Energy Efficiency of Milling Processes: A Review of some Fundamental Investigations and their Application to Mill Design.* Fourth European Symp. Comminution. Preprint 1–15

Starke, H. R. (1935): A Study on Ball Mills and Tube Mills. *Rock Products* **6** 40–46

Suzuki, S. (1955): Research on the Grinding Capacity of Open Circuit Cement Mills. *Proc. Japan Cement Eng. Ass.* **9** 189–184

Svensson, J. and Murkes, J. (1957): *An Empirical Relationship between Work Input and Particle-Size Distribution Before and After Grinding.* Int. Mineral Dressing Congress, Stockholm, 37–66

Taggart, A. F. (1945): Handbook of Mineral Dressing. John Wiley New York, 6–46

Takahashi, H. and Tsutsumi, K. (1972): Mechanochemical Effects on Zinc Oxide Powder Crystals. *Dechema Monographien* **57** 475–490

Tanaka, T. (1958): Bestimmung des Mahlmechanismus in typischen Trommelmühlen. *Staub* **18** 157–168

Tanaka, T. (1962): *Preferential Grinding Mechanism of Binary Mixtures.* Symp. Zerkleinern. Verlag Chemie Weinheim, VDI Verlag, Düsseldorf, 360–372

Tarján, G. (1972): Az ásványelőkészítés fejlődésének jelenlegi irányzata és jövője (Present Trends and Future of Mineral Dressing). (In Hungarian.) *BKL Bányászat* **105** 554–561

Tarján, G. (1974): Múlt, jelen és jövő az őrlés területén (Past, Present and Future in the Sphere of Grinding). (In Hungarian.) *BKL Bányászat* **107** 839–850

(Tovarov, V. V.) Товаров, В. В. (1957): Кинетика размола в барабанных мельницах. Труды совещания по прумению вибропомола в промышленности строительных материалов *(Kinetics of Grinding in Ball Mills.* Trans. Conf. Application of Vibration Milling in the Building Materials Industry), Промстройиздат, Москва, 90–101

Tromp, K. F. (1937): Neue Wege für die Beurteilung der Aufbereitung von Steinkohlen. *Glückauf* **73** 125–131, 151–156

Uggla, W. R. (1930): Quelque phénomènes d'ordre statique et dynamique concernant les broyeurs à boulets. *Revue des Matériaux de Construction* 447–453

Wasmuth, H. D. (1969): Bestimmung der Mahlbarkeit und des spezifischen Energiebedarfs bei der Mahlung von Erzen und Gesteinen mittels Bond-Test. *Aufbereitungs-Technik* 284–289

Zeisel, H. G. (1953): *Entwicklung eines Verfahrens zur Bestimmung der Mahlbarkeit.* Schriftenreihe der Zementindustrie, VDZ Düsseldorf, **14**

LIST OF THE AUTHOR'S PREVIOUS WORKS
RELATED TO THIS MONOGRAPH

1958 Zementvermahlung im geschlossenen Kreislauf. *Zement-Kalk-Gips* **11** 529–543

1960 Mahlverfahren, Kornaufbau und Festigkeitsverlauf verschiedener Zemente. *Zement-Kalk-Gips* **13** 419–424

1962 Le processus du broyage et son état d'équilibre. *Revue des Matériaux de Construction* **558** 73–82 **559** 115–121

Theorie und Technologie der Zementvermahlung. *Silikattechnik* **13** 115–123

1964 *Principles of Comminution.* A monograph. Akadémiai Kiadó, Budapest, 163 pp.

1965 Einige Frangen der Zementvermahlung. *Zement-Kalk-Gips* **18** 259–264

1966 Theoretical Study of the Grinding Process. *Acta Techn. Ac. Sci. Hung.* **56** 199–208. Co-author: Kiss, I.

1967 Strukturänderungen bei der Klinkervermahlung zu extremen Feinheiten. Zerkleinern, *Dechema Monographien* **57** 495–507 Co-author: Opoczky, L.

1969 Feinstmahlung von Zementklinker. *Zement-Kalk-Gips* **22** 541–546. Co-author: Opoczky L.

Feinmahlung von Zementklinkern mit Anwendung von Mahlhilfsmitteln. *Freiberger Forschungshefte* **A. 480** 67–75. Co-author: Opoczky, L.

1970 Die Gleichmässigkeitszahl der Kornverteilung des Mahlgutes. *Zement-Kalk-Gips* **23** 401–406

1972 Mahlanlagen mit mehreren Sichtern. VDZ Kongress '71. Bauverlag GmbH, Wiesbaden, 224–227. See also *Zement-Kalk-Gips* **25** 85–88

1973 Grinding Body Size and the Hardening of Cement. *Cement Technology* 47–56

Hindernisse und Grenzen der trockenen Feinstmahlung. *Silikattechnik* **24** 114–116

Sichtvorgang der Kreislaufzementmühlen. *Tonindustrie-Zeitung* **97** 116–118

Limit and "Efficiency" of Fine Grinding. *Acta Techn. Ac. Sci. Hung.* **75** 23–33

1974 Quelques résultats de recherches effectuées dans le domaine du broyage fin. *Revue des Matériaux de Construction* **689** 216–225

1975 Einiges über Mahlbarkeitsbestimmungen. *Zement-Kalk-Gips* **28** 325–330

Energy Saving Problem in Cement Grinding Process. Annual Meeting of Cement Association (Japan). Preprint 28–43

1976 Limit and Efficiency of Fine Grinding. *J. Res. Assoc. Powder Tech. Japan* 276–284 (in Japanese)

Fine Grinding and Agglomeration. *Cement Technology* 165–169, 199–205

Energy Saving Problem in Cement Grinding Process. *Funsai* (The Micromeritics) 64–73 (in Japanese)

1977 L'avenir des broyeurs à boulets et à galets. *Revue des Matériaux de Construction* **706** 163–167

Ist die Kugelmühle die Feinzerkleinerungsmaschine der Zukunft? *Aufbereitungs-Technik* 526–531

Zeitgemässe Mahlprobleme der Zementindustrie. *Wissentschaftliche Zeitschrift der Hochschule für Architektur und Bauwesen Weimar* **24** 305–312

1978 Ideas on the Perspective of Fine Grinding. *J. Soc. Powder Tech.* **15** 9–16 (in Japanese)

Bedeutung der Strahlmühlen in der Verfahrenstechnik. *Maschinenmarkt* **84** 1050–1053

SUBJECT INDEX

impact mills 102ff, 110
impulse law 102
imperfection 63

jet mills 111ff

Kick's theory 10
kinetics of grinding 25ff

lattice distortion 56
– energy 56, 58
lime grinding 131
liner wear 96
Loesche mill 100
logarithmic normal distribution 18

material mixtures 46
matrix method 71
mean value of distribution 21
mechanical activation 53
– air separators 119
mechanochemistry 51ff
median 63
"Microplex" separator 121
mill diagram 82
minipebs mill 73, 129
mode 21, 63
Mohs hardness 40
multichamber mill 88
mutual effects of particles 47

normalized distribution 60

ore grinding 131

parabolic trajectory 75
particle size analysis 17
– – distribution 17ff
perspectives in grinding 133ff
Pfeiffer roller mill 100
pharmaceuticals, grinding of 132
planetary ball mills 136
Polysius roller mill 100
porosity 41, 105
pottery grinding 131
practical fine grinding methods 73

rake classifiers 131
Rehbinder effect 52
retention time 81

ring motor 134
Rittinger's theory 10
rock salt 40
rod-peb mill 130
roller mills 100
Rosin–Rammler distribution function 18ff

selective grinding 46, 49, 96
separators 118ff
shock shatter process 138
shock wheel mills 104, 106
silicalcite 55
single particle breakage 13ff
size module 21
slippage of grinding charge 75, 81
Snyder mill 138
special impact mills 110
specific surface 19
speed regulated ball mills 80
standard deviation 21
supercitical speed 81
"Super micron" mill 111
surface energy 14, 58

tandem mill 117
tensioactive materials 51ff
"Terra number" 63
third theory 11
time of wear 91ff
tower mill 140
trajectories 74ff
tricalcium silicate 41
Tromp curves 63, 68
tumbling mills 74ff
turbidimetry 20

uniformity coefficient 21, 27, 30, 38, 127

very fine grinding 108ff
vibration mills 108
volume theory 10

Wagner method 20
wear of grinding bodies 91ff
wet grinding 114
wet process separators 131
work index 12

Zeisel test 37

150